Technoscience and Everyday Life

The Complex Simplicities of the Mundane

by
Mike Michael

Open University Press

Open University Press
McGraw-Hill Education
McGraw-Hill House
Shoppenhangers Road
Maidenhead
Berkshire
England
SL6 2QL
email: enquiries@openup.co.uk
world wide web: www.openup.co.uk

and Two Penn Plaza, New York, NY 10121–2289, USA

First published 2006

A catalogue record of this book is available from the British Library

ISBN 10: 0335 217 052 (PB) 0335 217 060 (HB)
      13: 978 0335 217 052 (PB) 978 0335 217 069 (HB)

Library of Congress Cataloguing-in-Publication Data
CIP data applied for

Typeset by YHT Ltd, London
Printed in Poland by OZGraf S.A.
www.polskabook.pl

Contents

Acknowledgements

My thanks go to those who have certainly contributed to whatever good things there are in this text. Thanks are due for the many conversations – not always directly pertinent, but invariably richly suggestive – with Andrew Barry, Lynda Birke, Rob Briner, Nik Brown, Simon Carter, Alan Collins, Susan Condor, Peta Cook, Miquel Domenech, Mariam Fraser, Mick Halewood, Lesley Henderson, Alan Irwin, Gavin Kendall, Celia Lury, Finn Olesen, Chris Todd, Cathy Waldby and Andrew Webster. Thanks too to the many colleagues who criticized and encouraged at various presentations, and to those postgraduate students – at Goldsmiths, Barcelona and Copenhagen – who were subjected to, and in turn challenged, my thinking on technoscience and everyday life. Goldsmiths College continues to inspire and enable in all sorts of ways – not least by providing a year's well-deserved study leave. I also owe a huge debt to the Centre for Social Change Research, School of Humanities and Human Services, Queensland University of Technology, who put me up as a fellow for two months – many thanks to Gavin and Trish Kendall for looking after me in Brisbane. Parts of this book draw on data derived from an Economic and Social Research Council-funded project (Xenotransplantation: Risk Identities and the Human/Nonhuman Interface – L218252044) conducted with Nik Brown. It goes without saying that my own everyday life informs the following pages; special thanks to Marios and Katerina Michael for reminding me of the peculiar lineages that feed into everyday life, and extra special thanks to Bethan, Nye and Yanna Rees for selflessly illustrating its unruliness.

1   Between technoscience and everyday life

Introduction

My Saturday newspaper tells me that in 2020, Britain, and by extension much of the rest of the western world, will be a very different place (the *Guardian*, 2020 Supplement, 18 September 2004). Everyday life will be transformed in many ways. In rich societies such as Britain, microchips will be ubiquitous, making hitherto 'dumb' materials such as bricks, medicine jars and milk cartons 'smart'. Medicine bottles will be able to detect when they were last used and how much medicine was taken, and will inform their users about their next dose. The use-by date stored on the microchips in milk cartons will be monitored by the refrigerator, which can place orders for fresh milk. These are some of the fruits of technoscience that will impact on the everyday life of the future. There are innumerable others that we hear about. Let us take nanotechnology (technology at a molecular scale) as a currently still fashionable example (at least at the time of writing). Among the applications that have been projected are included: nanobots – molecular 'robots' – that will seek out and repair damaged tissue or cells in our bodies; nanoparticles that can be incorporated into textiles that become lighter, more durable and 'self-cleaning'; carbon nanotubes that will allow super-light super-strength materials, making transport (including space flight) much less energy-hungry. These visions of the future world vie with less happy ones that are also the partial fruits of technoscience. Most obvious among these is global climate change. Over and above my newspaper's jokey scenarios in which palm trees will line, and crocodiles will inhabit, the River Cam in England, typically it is a future world of environmental havoc and devastation that is depicted. Of course, future technology will also serve to mitigate, to some degree, these negative technoscientific impacts; for example, we are told that nanotechnology will be used to produce ultra-efficient and cheap solar cells, thereby reducing the need for fossil fuels.

These are complex and contradictory views of the everyday life of the future. Contained within these forecasts are seeming shifts, more or less fundamental, in the ways in which we westerners comport ourselves. Our bodies become interwoven into information and communication networks, and they become subject to the circulation of biological materials and techniques that correct and enhance them – indeed we seem to be heading towards some sort of 'posthumanity'. In parallel, society becomes network society or biosociety, in which technoscience plays an ever more prominent, even revolutionary, part. As political subjects, we are at once increasingly encouraged to deliberate over these innovations even while they are presented as inevitable, or vital to our economic well-being. In this context of rapid change, our experience of space and time is challenged: space is multiplied through, complexified in and emergent out of an everyday that is crosscut by innumerable technoscientifically mediated connections; and time contorts as the future seems increasingly nearer as, with the aid of yet more technoscience, we rush ever more quickly towards it. In all this, our sense of self – embodied, social, political, located in recognizable space-and-time – undergoes some interesting shifts: we become at once more distributed and more singular.

Tacit in these predictions is a contrast between an exotic future and a mundane now. Current everyday life appears, by comparison to these strikingly different futures, boring, tedious and repetitive. There is an irony here. These futures are already being talked about now. Indeed, chronic to everyday life are forecasts, predictions, future scenarios in the newspapers, on the TV, in science fiction films. One might say that the technoscientifically novel, exotic, alien are part and parcel of the narrative baggage of our everyday lives.

Now, (if and) when these fruits of technoscience do eventually arrive, they enter a sociotechnical world populated by old technologies and established forms of action and perception. The promise of the new, of change and transformation, must be realized among 'old' technologies that are more or less integrated into everyday routines. In other words, it would seem that everyday life is characterized simultaneously by the promise of exotic technoscientific artefacts, the actual arrival of new technoscientific products and the mundane uses of existing technoscientific products – that is, by rapid change and quotidian repetition.

And yet, things are even more complicated. At the most obvious empirical level, novel technologies are routinely represented as old, or at least minor developments of existing technologies. Further, new technologies to which some controversy attaches become familiar through the increasingly pervasive efforts to inform or consult the relevant publics. Arguably, this is a part of what it means to be a 'technoscientific' citizen. Conversely, mundane technologies are not uncommonly promoted as new, or appropriated and put to novel uses (as well as sometimes disordering the everyday routines of which they are part). Here, then, we glimpse how the meanings of old and new, or exotic and mundane, are themselves very much up for grabs.

At a theoretical level, things are more complex too. Much has been assumed in the above use of the terms 'technoscience' and 'everyday life'. These are by no means simple terms, as we shall see. For the moment, we can point to the fact that both have divergent meanings in the academic literature that we shall later review in some detail. Suffice it to note that 'technoscience' can refer to the interwovenness of science and technology, or of knowledge and technique in which technology is indispensable for the production of scientific knowledge; alternatively, to study technoscience is to trace the multiple local processes by which expert knowledge is 'made' and comes to circulate and produce particular sorts of social and material orderings. 'Everyday life' can refer to the mundane, routine and generally unexciting character of social intercourse in which people's potential is by and large suppressed; on the other hand, to study everyday life is to unravel the subtle techniques by which social intercourse is managed and made seemingly seamless.

These perfunctory contrasts cannot disguise some intriguing points of theoretical intersection between everyday life and technoscience. If the 'critical' study of everyday life points to the ways in which such processes as consumption and spectacle disempower (and sometimes empower) people, the critical study of technoscience points to the way that the products of technoscience mediate relations of power in which certain groups, communities, collectivities are disadvantaged (and enabled). If the processual study of everyday life seeks to show how mundane social interaction proceeds, the processual study of technoscience examines how the social processes of everyday life also characterize the doings of scientists, both among themselves and in relation to such others as politicians and publics. Moreover, both 'technoscience' and 'everyday life' can be used 'methodologically', so to speak. It has often been suggested that to examine the unnoticed minutiae of everyday life is to pursue something like the essence of an epoch. In some approaches to the study of technoscience, to follow the engineer and the texts she sends out from her laboratory is to trace the social and material patterning of the world. In both cases, then, studying the apparently small, local and insignificant allows us to derive insights into the apparently big, global and important. However, as we shall see, even 'the big' and 'the small' are problematic terms. After all, the 'smallness' of everyday life can be regarded as infused with such 'big' contemporary dynamics as globalization and individualization. At the same time, these 'big' dynamics get instantiated only in the 'smallness' of everyday lives, and, in any case, are articulated as 'big dynamics' in the 'small' locals that comprise the everyday practices of, for example, economists and sociologists.

The vicissitudes of velcro

Let us a take a moment to reflect here. These introductory remarks have touched upon the complex and contestable dichotomies of old and new, big and small, process and critique, change and repetition. Hints have been dropped about the roles that technoscience plays in the ways in which bodies, politics, societies, times and spaces are made manifest in everyday life. This is indeed a broad sweep of issues, concerns and topics.

Yet this sweep hardly captures all the connections between everyday life and technoscience. To get a better and more concrete, though certainly not comprehensive, view of this multitude of links, let us take as an example a mundane technoscientific product whose operation in everyday life is as unremarkable as it is widespread – velcro.[1] Here is an artefact that incorporates the ordinariness of everyday life while also being manifestly the product of technoscience. Yet as we explore it in its various manifestations – as a bit of material culture and a representation that circulates – we happen upon a morass of meanings, relations and effects.

Velcro® is a brand name for a form of hook and loop fastening. There is an origin story attached to the invention. In 1948 Swiss engineer George de Mestral, after a walk, became intrigued by the mechanism by which cockleburs (a plant related to the sunflower) attached to his clothes and his dog. Studying them under the microscope, he noted that cockleburs were covered by thin strands with hooks on the ends that clung to the threads that made up the fabric of his clothes. The traditional 'a-ha moment' came when he 'recognized the potential for a practical new fastener' as one account put it (http://www.ideafinder.com/history/inventions/story015.htm, accessed 5 July 2004). Over the course of eight years de Mestral perfected the hook and loop fastening, which comprises two strips of nylon fabric, one side consisting of thousands of small stiff hooks, its opposite made up of many tiny loops. These mimic the action of the cocklebur – the hooks fasten onto the looped plastic threads to produce a very strong bond. The term 'velcro' was derived from the French for velvet (velour) and hook (crochet). A particular advantage of this fastening technology over something like the zip is its ease of use – no precise alignment is necessary. This origin story is a commonplace (in the rhetorical sense – see Brannigan 1981): it tells about the moment of inspiration and the years of perspiration; it tells about the way nature provides the technical principle – a principle that can be transported across domains (and functions); it even hints at the artistry of naming – or what would nowadays be called 'branding'. What we see here is a particular romantic version of the technoscientific innovation process in which the figure of the scientist as hero looms large. This contrasts greatly to the Velcro® company website where continued innovation appears routinized and collectivized.

On visiting the Velcro® company website, the range of hook and loop products appears enormous. There are three main headings: Automotive, Industrial and Consumer. Under 'Consumer' we have subheadings such as Sewing and Home Decor, The Great Outdoors, Hardware–DIY, Party and Holiday, Office/School, and Beauty. Under 'Automotive', products are available for seating, electrical systems, overhead systems, trunk storage. For 'Industrial' there are applications that can be used from 'hometown grocery store to the space shuttle', and products ranging from baby seats to medical braces. All these products work along the principle of fastening two strips of plastic together. Depending on what these two strips are themselves attached to, two different functions are possible. Attached to each other by some material or other, they can serve as containers of some sort. Thus we have packages – plastic packs that are sealed along the top with Velcro® rather than some adhesive or zip or slider. We also have wraps of one sort or another – more or less thin strips of material or plastic topped and tailed with hook and loop Velcro®, which join together to form a ring of plastic that can, for illustration, be used to gather up and tidy cables or wires in the home, office or in car assembly. Velcro® is also sold as strips or coins that can be attached to garments as fastenings that replace buttons or zips. In all this, Velcro® functions to enable some form of containment (though this can be turned on its head to become a technology for quick decontainment or release – recall the male strippers of *The Full Monty*). Here, then, is the other overarching function of Velcro®: the attachment of two separate surfaces. Strips or coins of Velcro® (with adhesive sold separately or already incorporated) can be used for displays, as alternatives to hooks for securing things (e.g. tools) to other surfaces like walls, or for attaching cushions to car seats.

In all this, we see that velcro, by virtue of its basic principle of operation, meshes with many other technoscientific artefacts that are found across a wide range of activities – leisure, work, communications, domestic, travel, medical, scientific. There are two initial points to note here. First, velcro evokes the enormous multiplicity of ways in which technoscience mediates our everyday life. Though this is an obvious enough observation, it does raise a number of issues about how one goes about tackling a 'topic' such as 'technoscience and everyday life'. This will be addressed in the next section. Second, the example of velcro illustrates that technoscientific artefacts should not be uncritically regarded as 'singularities' – stand-alone unitary entities: rather, they are tied into, and function in relation to, complexes of technologies, knowledges and cultures – what we shall later call 'sociotechnical assemblages'. Indeed, without the appropriate context such technologies would not 'work'. From this second point, we can derive two additional, and again contestable, dichotomies: singularity and relationality (sometimes velcro might be best treated as a singular artefact, sometimes as an entity that emerges out of a range of relations), and visible and invisible (sometimes velcro operates behind the scenes, sometimes its presence is highly salient).

Now, one obvious signification from the range of uses to which velcro is put is that it is 'unproblematic' – it is easy to use, in multiple contexts, and there are few things that go wrong with it. As it says on the Velcro® site in relation to Velcro® packages: 'designed and engineered for all ages and motor skill levels' (http://www.velcro.com/industrial/touchseal.html).

So, it is an artefact avowedly immune to the differences in people's bodily capacities. Indeed, it would seem that velcro has been elevated and idealized as a principle of unproblematic attachment or fastening. However, as we shall see below, things are not so simple and in the specificity of its functions, velcro entails many assumptions about bodies and settings.

To the extent that velcro is an embodiment of a principle of unproblematic attachment or fastening, it is an emblem that is appropriated to signify many instances of such unproblematic fastening. For example, one potential nanotechnological innovation has been described as 'nano-velcro' in which carbon nanotubes intertwine (in actuality this entails hook–hook as opposed to the hook–loop connections of velcro). As such, velcro as an everyday object is used to popularize a technoscientific innovation. At the other end of the ontological scale, we find velcro has been used to describe social bonds and networks where contacts accumulate by virtue of a social stickiness likened to that of velcro (http://socialsoftware.weblogsinc.com/entry/5326667344598756/). In this case, velcro as a technoscientific artefact is deployed to capture a particular social phenomenon. I have deliberately overstated these divergent uses of the representations of velcro in order to highlight the two-way traffic of representations between everyday life (characterized as 'lay') and the realms of technoscience (characterized as 'expert'). Indeed, technoscience and everyday life borrow from each other to the point that, as various authors have argued (e.g. Martin 1998; Irwin and Michael 2003), it is perhaps better to focus on the circulation of shared representations than on the differences between expert and lay domains. For example, the biomedical depiction of the immune system draws on representations current in everyday life: as Martin (1994) traces, within biomedicine the shift from military metaphors to those where the immune system was seen to be flexibly responsive 'shadowed' broader social (and economic) changes.

Now, for want of a better term, the 'velcro principle' as a means to understanding the stickiness of social relations and the ways in which we collect social contacts might be seen to be rather spurious. Nevertheless, it does hint at the various ways in which technoscience is used to illuminate the processes of everyday life. For example, there appears to be an emerging genre of books that aim to explain everyday phenomena as well as popularize science, by applying particular scientific techniques to mundane practices. Thus we find such volumes as *Why do Buses come in Threes: The Hidden Mathematics of Everyday Life*, *How Long is a Piece of String?: More Hidden Mathematics of Everyday Life*, *How to Dunk a Doughnut: The Science of Everyday Life*, or *Mind Wide Open: Your Brain and the Neuroscience of Everyday Life*. This genre will not be examined in any detail here, but at the very least it illustrates yet another layer in the relation between technoscience and everyday life in which, on the one hand, the content of everyday life can be 'better' understood through science and, on the other, everyday life as a process can generally be enhanced by the assimilation of scientific knowledge. Add to this genre the innumerable popular accounts, across the whole range of media, of scientific discoveries and technological inventions, and we have an enormous outpouring of technoscientific popularization. The relation between texts and audience is generally a very difficult one to disentangle, but we can minimally note that whatever complex versions of lay reader are entailed in such texts – whether consumer, citizen, parent, cultured person, and so on and so forth – none would fail to be improved by having contact with the verities of technoscience. Less flippantly, we might say that many such identities are partly mediated through the representations of popular(izing) technoscience. In other words, part of the discursive resources we have available to us in everyday life – resources through which we articulate and enact ourselves – are derived from technoscience. The irony here is that these technoscientific resources are themselves, as we have noted above, partly derived from representations that circulate in everyday life.

In the preceding paragraphs, the focus has been on how the representation of velcro (and more broadly technoscience) affects self-identity. However, velcro is of course a material artefact encountered in everyday life in more or less noticeable ways. As mentioned above, velcro often works 'invisibly' – quietly doing its job behind the scenes (for example, by attaching seats to their frames in public transport), or 'automatically' when we use it directly, when we tighten the straps on our sandals, say. Yet, we also noted that for velcro to 'work' certain conditions need to be in place. These become particularly apparent when things go awry.

Despite the impression of infallibility, velcro does go wrong: it gets too loose and it gets too tight. Regarding the former, practitioners of cyclocross – a sport involving cycle racing on short off-road courses, which involves the use of racing-style bikes that are ridden, carried, bunny-hopped and skidded in the course of the race – have noticed that the velcro straps on their shoes loosen up in wet and muddy conditions. Anecdotally, it is not uncommon to find that velcro used in the proximity of fluffy material tends to collect lint and loosen up, necessitating users to pick or brush the fibres out in order to get the velcro once more to operate relatively normally. Regarding the problem of too much adhesion, it has been reported that prolonged use of velcro can make it difficult to open – this is particularly worrisome if the velcro covers the reserve parachute pins (the solution is to open and close the velcro regularly before each flight in order to reduce its resistance – (see http://parapente.para2000.free.fr/wings/safety/note_kortel-harness.html, accessed 2 June 2004). Sometimes, other problems arise. For example, we find complaints that velcro fastenings on Rikki wraps – plastic wraps that go over cloth nappies (diapers) – cause discomfort. One mother complains on an electronic bulletin board that 'lately the velcro starts rubbing in my son's chubby thighs as he crawls and cruises and therefore leaving red marks there'. The advice she receives is that this is a common problem that is easily solved by ensuring the 'hooks are relatively lint free. This way the velcro can stick best.' Further, it is usually best, and especially so for the particular product under discussion (Rikki wraps), that the hook and loop side of the velcro attach perpendicularly ('completely straight on', see http://www.mother-ease.com/wwwboard/messages/1065.html). In all these cases, the solution requires changes in bodily comportment. The use of the body must be altered in various ways in order to allow velcro to work: avoid mud, pick out lint, repeatedly open and close the velcro attachment, ensure the alignment of velcro strips. The obverse of this is that certain assumptions are built into the functioning of velcro. By way of a prelude to the discussion in Chapter 2, we might say, albeit crudely, that technoscientific artefacts exercise a certain 'power' over us in that they necessitate certain bodies and bodily practices. Needless to say, this can be resisted in various ways – from making different consumption choices such as buying cyclocross shoes with ski boot-type fastenings to overt political interventions (see below).

Velcro as a technology that operates behind the scenes can also 'go wrong'. For example, in Edinburgh, the Bus and Coach Watchdog has produced a report alleging that, because seat cushions and backs are attached to seat frames by inadequate velcro strips, passengers are exposed to unnecessary risk. These cushions and backs are often used by passengers to support themselves as they get up or try to pass by other passengers. This is especially the case at the rear seating, where passengers must also negotiate a steep step and deal with the fact that the steadying poles are not within ready reach. As the report put it: 'People relying on the seats being secure could find themselves thrown forward as the cushion comes away' (Bus and Coach Watchdog 2003). Here the response, embodied in the report, is to apply pressure to local transport authorities to ensure the safe attachment of cushions to seat frames in buses. The point here is that technoscientific artefacts can entail dramatically 'inappropriate' assumptions to which the only response is political (in the usual sense). Of course such 'inappropriate' uses of velcro might well reflect all sorts of conditions – economic, technical, regulatory – which might be used to justify such uses. However, at base, the argument over the appropriateness of such uses is a contest over the nature of the local sociotechnical world: who can be expected to do what with what. The accessibility of steadying poles becomes key here. If the bus company insists that it has provided enough steadying poles and that all passengers can be expected to access these, then the arguments centre around the question of 'whose reality?' As we shall see, these are realities that are further complicated by the form such arguments take.

These examples of things going wrong underscore once again the complexity of the relation between technoscience and everyday life. Indeed, as we shall see, 'going wrong' is one means by which innovation arises in everyday life. Interestingly, velcro also points to the way that, even when things go particularly well, they can go wrong. In other words, it is the very unproblematic-ness of velcro that renders it, for some, problematic. Let us take the example of a religious piece from the Christian Reformed Church of North America (i.e. a sermon). The argument here is that the ease of velcro has robbed children of a 'great learning experience'. Children miss the sense of achievement of learning to tie laces. However, crucially, they miss out even more by being denied the opportunity to untie laces. To quote: 'Velcro . . . also shields them [young children] from the little problems that can crop up if you untie your shoes carelessly . . . you form a nasty little knot that exasperates you as you try to undo the damage.' This is not trivial because, as the writer puts it, 'we learn about the big things in life by paying attention to the little things' – something certain species of microsociologist, sociologist of every-day life, critical theorist and sociologist of scientific knowledge would be happy to endorse. The lesson from this is that such little activities and the difficulties they throw up enable children to learn how to learn; as such, the convenience of velcro denies them one such opportunity (http://www.backtogod.net/sermons/sermon_deail.cfm?ID=36060, accessed 3 June 2004). However we might view this particular criticism of velcro, this essay does nevertheless engage with what it means to lead the 'good life', and the role of such technoscientific artefacts in the imagining and enablement of such a 'good life'. This brings into focus another dimension of the relation between technoscience and everyday life – their alignment or otherwise with the fu-ture, with the ebb and flow of what used to be called progress. But such discourses – in this case the 'use of laces' is a plea for a better past to be resurrected in the present in order to produce a better future – are about the making of the present, they are performative. The discourses, and sometimes spectacles, that 'accompany' technoscience into everyday life serve in the reordering of the present in order to fashion a future everyday life.

In this section, we have explored, through velcro, a variety of ways in which 'technoscience' and 'everyday life' intersect. Indeed, when one multi-plies these intersections by the myriads of other intersections that are co-present even in these episodes, one is tempted to ask the following question: 'What is not everyday life and technoscience?'

How to navigate the morass

In this introduction, I have attempted to give some flavour of the variety of intertwinements of everyday life and technoscience. At the very least I hope I have illustrated some of the ways that everyday life and technoscience are crucial to one another empirically: everyday life is permeated by tech-noscientific artefacts, by projections of technoscientific futures and by technoscientific accounts of the present. Technoscience plays a pivotal role in the way bodies, identities, society, citizenship, space and time are articulated – in the dual sense of being talked about (discoursed) and practically realized (enacted) – in everyday life. It has also been suggested that at the conceptual and theoretical level, studies of everyday life intersect in a number of ways with the analysis of technoscience. Indeed, these intersections have been loosely structured around a series of heuristic, though nevertheless dubious (because contestable and collapsible), dichotomies: old and new, big and small, process and critique, change and repetition, singularity and relationality, and visible and invisible. Putting together this intertwinement and these intersections, the pervasive sense is that we have entered an empirical and theoretical morass. Moreover this is a morass that is deepened when one takes into account the view that the everyday life of the academic analyst is a setting in which, through certain tools – not least technoscientific ones – versions of everyday life and technoscience are fashioned and circulated.

Navigating this morass will hardly be easy. Certainly, there are ways of rendering the process of navigation easier – that is, of making the narrative movement of the present volume manageable. For example, one could focus on 'key thinkers': one could list a number of writers around whose work one could usefully structure discussion about the connections of everyday life and technoscience; excellent models for this sort of approach can be found in Gardiner (2000) and Highmore (2002). For instance, chapters on Goffman, Lefebvre and De Certeau might respectively consider how technoscience features in the presentation of self, or in the dialectic of alienation and 'presence', or in the interweaving of strategies and tactics. Alternatively, one could pick out a range of 'areas of everyday life' and explore the complex role of technoscience in these – an approach exemplified in Bennett and Watson (2002) and Silva and Bennett (2004). For instance, we might look at the roles of technoscience in such domains as home, travel, work, the street, shopping, the family, and so on. Inevitably, each of these arrangements raises problems: which theorist do we include and which do we neglect? How do we differentiate these everyday domains and what do we do about the routine transition from one to the other? The organizational strategy adopted here is no more or less satisfactory, though it does have the advantage of addressing the issues that most interest me.

After a chapter that sets out some of the relations between technoscience and everyday life within the context of a number of relevant literatures, the rest of the book will be organized around a series of thematic constructs: the body, citizenship, society, space and time, and finally, drawing all these together, identity. Partly these have been chosen because they appear so crucial to contemporary social science; partly they have been chosen because they allow us to interrogate some of the rich interweavings of technoscience and everyday life in relation to a range of empirical settings. Initially, my thought was to have each of these chapters illustrated with the same examples (e.g.

walking down the street, cooking and eating, taking a shower and getting dressed, working) in order not only to provide continuity across chapters, but also to allow the complexity of the relations between everyday life and technoscience to unfold. However, it soon became apparent that this made the text rather repetitive and boring (which is not to say it's not still boring; if it is still boring, it's in a way I find agreeable). Instead, I opted for presenting a range of disparate examples across the chapters: the use of Post-it® Notes in the mediation of relations of power, the practicalities of sleep in the complex enactment of bodies, the micropolitics around traffic surveillance cameras, the role of the mobile phone in the production of spaces, the misuse of the tape recorder in revealing everyday society, the arguments over xeno-transplantation in the performance of futures, the technologies of queuing in the mediation of multiple temporalities. There are some examples that crop up on a regular basis (e.g. the promise of xenotransplantation) and these partly serve to connect the chapters. The issue of connection is directly addressed in the final chapter, where bodies, society, politics, space and time are brought together in a discussion of identity, technoscience and everyday life. Another commonality across chapters is that each is structured by the tension between the 'old' and the 'new', the novel and the mundane. For example, walking down the street entails 'old' technologies such as shoes and paving stones, but also 'new' technologies such CCTV, mobile phones and 'smog detectors'. Each of these has different implications for the way the body is used, how space and time are enacted, how citizenship is performed and how 'society' is mediated.

Inevitably, the relation between technoscience and everyday life is so multifarious that the present volume can only scratch the surface. Thus, there are several broad sociological concerns and empirical domains that fail to make an appearance here, or else are treated only flimsily. These reflect my many biases, my few interests and my proliferating ignorances. To give an indication of what has 'gone missing', I can point to the following. Empirically, I do not to deal with, for example, the role of finance, romance or education in everyday life, and where I deal with work settings these tend to be offices rather than the factory floor or the retail outlet. Also fallen by the wayside is a detailed consideration of the putative 'others' of technoscience, those more or less everyday knowledges and practices – such as astrology, herbalism or shamanism – that are routinely contrasted to the knowledges and practices of technoscience. In terms of theory, while I address several writers on everyday life, I tend to concentrate on Lefebvre and De Certeau to inform many of my discussions. In engaging with relations of power, I address gender (or rather gendering), but not 'race' or age or disability or sexuality and the technoscientific arrays that are instrumental in rendering these. In relation to technoscience, there is, perhaps, a slight empirical leaning towards biomedical science and artefacts, though I do also address a variety of information and communication technologies. When discussing the future or exotic character of technoscience, I don't especially attend to the role of science fiction, though I do touch upon this genre. 'Finally', my theoretical perspectives on technoscience tend to come out of what might be termed the microsociological schools and, as such, I don't make much mention of those approaches – philosophical or sociological – that are informed by critical theoretical, postcolonial or feminist writings (e.g. Winner 1988; Hess 1995; Feenberg 1999). Such limitations are unavoidable, and the fond hope is that the absences, as much as the presences, will make what is to follow novel and engaging.

This book

If the foregoing has laid out some of the reasons why 'technoscience and everyday life' – theoretically and empirically – might be of interest, the rest of this chapter sets out how such interest is manifested across a range of 'thematic constructs' as I have called them above. Chapter 2 introduces the two main literatures that inform the present volume: the sociology of everyday life, and the relevant parts of the sociology of science and technology. I deal with the former by extracting a number of what seem to me to be key foci: the theorization of everyday life as taken-for-granted, or as elusive, or as 'bad' (in the sense of entailing repetition, discipline or alienation), or as 'good' (in the sense of enabling critique, promise or resistance). With regards to the sociology of science and technology, I draw on those approaches that treat technoscience as a circulation that 'ties together' the human and nonhuman in particular ambiguous or multiple ways. For both literatures, my theoretical emphasis lies with the 'micro'. These literatures are supplemented by others, notably those that, albeit unsatisfactorily, can be termed 'technoculture', 'material culture', 'domesticating technology' and 'scientific citizenship studies'.

Chapter 3 explores the body in relation to technoscience and everyday life, showing how it is at once 'opened up' and 'closed down'. The latter connotes the non-fragmentariness of the body, and this is shown to be partly dependent on the former in which the body's multiplicity is enacted with and through the complex circulations of technoscience. While aspects of this multiplicity rest on such mundane artefacts as electric toothbrushes, mattresses and vacuum cleaners, others derive from the more esoteric, posthuman reaches of technoscience (which nevertheless circulate in popular culture). Chapter 4 considers 'politics', primarily in two guises: everyday 'micropolitical' routines around issues of technoscientifically mediated surveillance, and the 'making' of technoscientific citizens through the popularization of participatory mechanisms in scientific decision and policy making. With regard to the surveillance, I tentatively propose the notion of the taleidoscopticon (to contrast with those of panopticon and oligopticon) as a way of evoking the complex mutualisms of surveillance as surveillers are themselves surveilled. In the case of the 'technoscientific citizen', it is suggested that, however much this figure might contribute to institutional decision making, such decision making is faced with chronic uncertainty that is, ironically, 'resolved' through borrowings from everyday performances of 'authenticity'. Both of these political 'moments' reflect, in their different ways, the problems raised in the pursuit of 'transparency'. An integral part of politics is, of course, the use of rhetoric, and 'society (is such and such)' is a ready-to-hand part of mundane political discourse. And yet it can be contended that everyday 'social' life is not possible without the mediations of technoscience.

In Chapter 5, this heterogeneity of society will be exemplified by critical analyses of everyday situations regarded as 'social', including the production of ostensibly 'social' data in sociological research practice. It is the largely unnoticed operation of mundane technologies – such as tape recorders, Post-it® Notes, mobile phones – that enables the enactments of particular versions of 'society'. At the same time, 'society' has been subjected to the 'encroachment' of new technoscientific developments (iconically, biotechnology and information and communication technologies) that have, arguably, drastically changed current conceptions of society. In addition to considering the performativity of the accounts of these emerging societies (especially virtual society), I attempt to tackle how we might grasp the multiplicity of 'new' technoscientific societies in circulation. So, in addition to virtual, we might list, for example, geneticized and surveillance societies: the question posed is 'What is the effect of this proliferation of societies?' Chapter 5 finishes with an interrogation of the successful police drama TV series, *Crime Scene Investigation*, in order to unravel some of the assumptions made about the society of technoscience – that is, how science is 'done' within (or, more accurately, without) society. Chapters 6 and 7 consider technoscience and everyday life in the context of the production of, respectively, spatiality and temporality. Of course, it has not escaped attention that space and time are nowadays thought to be integral to one another – space-time is commonplace to the sciences, is beginning to be properly theorized in the social sciences (e.g. Jones, McClean and Quattrone 2004) and is increasingly accessible to common sense. It will be taken as read that, although I will focus my discussions primarily around either temporality or spatiality, each term presupposes the other. In investigating some of the roles that technoscience plays in the emergence of spatiality in everyday life, I consider a number of more or less mundane technologies – mobile phones, maps, sunglasses, Teslar watches. What interests me here is the way that certain spatial divides are produced and moralized or politicized. In this context I am keen to examine some of the intricacies of the relation between the micro and the macro, or the local and global, and to trace the contestations and calibrations of the local, multisensorial experience of everyday spaces and technoscientifically mediated renditions of 'those' spaces. I also address how spatiality unfolds in the relations between entities and how the entities emerge out of those spatialities. In considering 'clutter' and 'roadkill' I attempt to think a little about the sorts of prepositions that might be useful in developing this sense of emergent and multiple spatiality. In Chapter 7, temporality is likewise treated as emergent and multiple. On this score, the Tour de France triumph of Lance Armstrong (and its everyday reportage) proves to be a potent illustration of the complexities of time. The rhetorical dimension of time – particularly in the form of the construction of the future – is taken up in terms of 'temporal performativities'; how these operate to enact futures in, and of, the present is considered in relation to xenotransplantation. The chapter also explores the more general issue of the 'impacts' of 'mundane' and 'exotic' technoscience – both to 'remake sameness' and 'effect change'. This is theorized through the idea of the complex patternings of ordering and disordering out of which emerges temporality (or multiple temporalities). The 'queue' (specifically, the complex queuing system at a railway station) is used to illustrate some of these processes in the concrete. Finally, I consider the way that temporality might be linked to future prospects, and the hope that might be invested in these.

Electric toothbrushes, mattresses, vacuum cleaners, technoscientifically mediated surveillance, mechanisms of public engagement for technoscientific decision making, tape recorders, Post-it® Notes, mobile phones, maps, sunglasses, Teslar watches, 'clutter' and 'roadkill', cycle racing, xenotransplantation, queuing systems – these are some of the examples that will be found in what is to follow. The final chapter asks what it is to think of identity in relation to this everyday technoscientific multitude – a multitude that encompasses, at the very least, seemingly singular items, complex events and sophisticated systems. I restate the present commitment to the 'micro', and briefly show how the examples used to illustrate the various themes can be rethought in terms of the other themes covered in this book. A queue is thus not just about temporality but spatiality, society, politics and bodies – and identity. In fact, all these are indissoluble. I return to the example with which I began, namely velcro, in order to unravel a few of its identities, and unpick some of the correlative everyday social, and hybrid, identities. Inevitably, these turn out to be complex, relational, heterogeneous, emergent and multiple.

Note

1. Velcro has become, much like hoover and the vacuum cleaner, synonymous with hook and loop fasteners. In this book, I use 'velcro' in this way, unless referring to the actual branded product, in which case I use 'Velcro®'.

2 Versions of everyday life and technoscience

Introduction

In Chapter 1, I attempted to establish and illustrate the mutual interwovenness of everyday life and technoscience, and while I hinted that the intersections were many, varied and complex, I was careful to be cautious about such intersections. This is because how one grasps the nature and significance of these intersections rests on the sort of theoretical tradition out of which one emerges and the conceptual armoury one deploys. Behind the examples littering Chapter 1 are a variety of approaches, perspectives and interests that make the notions of 'everyday life' and 'technoscience' at once so compelling and so slippery. The present chapter is thus concerned with surveying[1] the relevant literatures that address, respectively, everyday life, technoscience and technoscience-and-everyday life. However, as the last of these suggests, these literatures overlap at various points which, of course, makes the process of surveying less than straightforward, not least when it comes to the organizing principle behind the surveying.

Now, one obvious way of representing the literature is through a comparison of the main traditions (e.g. Bovone 1989) or the key authors in the field. There are recent excellent volumes that do the latter, importing their own distinctive criteria or dimensions of comparison. Highmore (2002), for example, is particularly interested in the difficulties of representing 'everyday life' given that it is at once immediately accessible and profoundly elusive, and, indeed, realized in both the material and the virtual (see Seigworth 2000). He thus compares some of the various techniques by which the everyday has been approached from the massive and heterogeneous data collection of Mass Observation, to the montage stylistics of Simmel, Benjamin and the surrealists. Gardiner (2000), by comparison, compares authors (and traditions or movements) in terms of how they articulate the possibility that within everyday life – for all its routine-ness, repetition, alienation – one finds (or implants) the seed of 'resistance' or 'transcendence'. As such Gardiner unravels the commonalities and differences between, for example, Lefebvre's idea that everyday life can entertain 'moments' in which it is possible to glimpse the world transformed (however diffuse that vision might be) and De Certeau's notion of tactics in which subtle resistances that invoke the past and recreate the present momentarily rupture the strategies of discipline.

In initially contrasting Highmore and Gardiner, we face two rather different ways of comparing what are seen to be the founding analysts of everyday life, respectively, in relation to the issue of the representation of everyday life, and the possibility of the 'transcendence' of everyday life. By comparison, in this chapter, I intend to organize my survey in terms of different criteria of comparison. For instance, many authors point to such qualities of everyday life as its 'static' or 'repetitive' nature; some attempt to grasp and represent its fragmentary or piecemeal character; others identify it in terms of its 'basic-ness' or its status as a deep, presupposed 'background' that structures social activity; some seek out the capacity of everyday life to resist mundane impositions, or to inspire an immanent apprehension of a world in which 'life is better' (to put it crudely). These themes, already touched upon in Chapter 1, are abstracted from the literature and here serve as heuristics (to do them proper justice would require several volumes) through which to anchor my survey of the literature.

When we turn our attention to technoscience, the literature is no less fraught with difference. Technoscience comes in various versions. It can be seen as the embodiment of the contemporary indissoluble interdependences of science and technology; it can be regarded as a process of circulation of various artefacts and texts that mediate and translate the 'world'; it can be seen as prime medium for the construction and emergence of complex and heterogeneous figures – such as the cyborg – which, though epistemologically, ontologically and politically ambiguous, can profoundly shape our understandings of, and relations to, nature and society. This literature will likewise be surveyed, paying particular attention to what it is that might *count* as technoscience, both formally (as a process of world making) and substantively (as more or less specific procedures, products or configurations).

The two surveys of, respectively, everyday life and technoscience will make up the next two sections of this chapter. The following section addresses a ragbag of literatures that, in one way of another, interrelate technoscience and everyday life. Such interrelations entail, for example, the role of scientific knowledge in enabling people to be 'good citizens', the idea of material culture where material artefacts (many of which can readily be brought under the rubric of technoscience – e.g. cars, televisions, telephones) are seen to become part and parcel of everyday routines, or the concerns with technoculture where motifs and metaphors derived from science, science fiction and everyday life influence the ways in which people habitually come to apprehend both particular innovations and the processes of innovation itself. Rather than surveying, my approach in this case will be more peripatetic (Michael 2000a) – a gentle meander through, and more or less haphazard encounters in, a forest of intellectual traditions.

In the final section, in light of this peripatetic, I revisit the themes of everyday life and of technoscience and, albeit briefly, reread them through one another. To be sure this is a circuitous route, but it has the merit, I hope, of taking the reader further into the intricacies of that innocent couplet, technoscience and everyday life. My aim in all this is simple: to signal the complexities of what is to follow in the rest of the book.

'Capturing' the themes of everyday life

Before we can address the various themes that might be found in and around considerations of 'everyday life', we should briefly address the interconnected issues of what 'everyday life' is (or what it is contrasted against) and why 'everyday life' is of interest now. Clearly both issues relate to the themes presented below, after all what counts as knowable as everyday life relates to a particular history that has at once to enable the emergence of the 'domain of everyday life' and to constitute that domain as an 'object of study'. Attending to 'history' first, for an author such as Lefebvre (1947), it is the differentiation wrought by capitalism in which everyday life is increasingly divorced from other spheres, especially work, that renders a particular everyday life in which creative human capacity, constrained by rigid roles and niches, is routinized and commodified. In contrast, for those authors who take a view of everyday life as taken-for-granted, or regard it in terms of fundamental social interaction or sociality (e.g. Schutz 1967), there is no need to have a history of 'everyday life' that traces its emergence; within this perspective, everyday life is simply a constitutive part of what it means to be a human social being. Nevertheless, one can still embed the emergence of 'everyday life' as a topic within a social, cultural and intellectual history. Thus, for instance, Bovone (1989) suggests that it is to the 'crisis in totalizing classical sociologies' (p. 42) that the recent interest in everyday life might be linked (certainly one can associate the emergence of ethnomethodology in the 1950s and 1960s with an exasperation with the structural functionalist model of social actors as 'dopes' – see Heritage 1984). Silva and Bennett (2004, p. 1) echo this sentiment when they suggest that it is partly because of scepticism that 'institutions and structures are timeless, fixed entities' but rather the product of 'active processes of human creation through ordinary interaction' that 'the category of everyday life is enjoying something of a renaissance in contemporary social thought'.[2] However, they also link this to broader dynamics, not least the deep everyday uncertainties so characteristic of late modernity (e.g. Giddens 1991). The simple point here is that what is to count as everyday life is partly contingent upon the particular history more or less explicitly assumed by the account of everyday life, whether that be the history of social thought or of the social world.

Now, we might also note that the academic writing of the everyday itself is embedded within the everyday dynamics that comprise academic life. As such, academic discourse (including discourse on everyday life) shares with everyday life the requirement to be accountable. This is a broadly ethno-methodological point (see below) in so far as it is based on a view that in-herent to social action is the making of such action warrantable. As social action, what is academically written and presented (and that incorporates empirical and theoretical dimensions) on everyday life has to be warrantable within academe. The previous paragraph points to the way that an aspect of this warranting is the evocation of a particular intellectual or social history. However, there are also other warrants available. For instance, analytic or critical writing on everyday life is tacitly contrasted with everyday life itself: it is heroic (tragic even) as opposed to mundane, or critical as opposed to rou-tine, or an 'intervention' as opposed to passive (see Gouldner 1975; Feath-erstone 1992). Yet such contrast – such academic heroism – is part of the everyday currency of academia: it is how certain critical writing is warranted. That is to say, the warrant for critical academic writing on everyday life rests, ironically, on an everyday differentiation from the everyday. As we shall see, things are still more complex as critical writers also detect immanent pos-sibilities and marvels in the everyday. Nevertheless, it is critical writers who 'excavate' these – the 'heroic' dimension of writing remains. This is not to somehow debunk such critical writing. Rather, it is to situate it within its own everyday and to underline Bennett's (2004) argument (see below) that wari-ness needs to be exercised in so far as contrasts that point to or enact 'others' – the hero, the critic, the activist, the creator – can serve to reduce the com-plexities and heterogeneities of the everyday.

The everyday as the taken-for-granted order

According to phenomenological sociology, everyday life can be said to be comprised of intersubjectively shared meanings. While there are individual differences in the perception of objects in the world, it is usually assumed – there is a 'natural attitude' in which – our individual standpoint can be interchanged with that of others and that any differences of perspective are pragmatically irrelevant (see Schutz 1967). Yet the common sense of everyday life might be said to be fragmented when one considers the multiplicity of social settings that now make up society. Thus, everyday commonsense thought that is habitual and unreflexive can be contrasted to more symbolic or specialist forms of thought such as that of science (Berger and Luckmann 1966; Bovone 1989, p. 45).

However, as ethnomethodologists and many others note, while the content of such knowledge domains and related activities might vary, their basic social *form* is much the same. There still remains a 'natural attitude' within each domain in which common assumptions are made about mutual perceptions, for example. Moreover, as ethnomethodologists note, there are basic features of social interaction that enable it to proceed (even when it is ostensibly disordered) in an orderly manner. In other words, in everyday life it is taken for granted that order will be sustained, or else resumed in an orderly fashion. Thus, as in Garfinkel's (1967) famous breaching experiments, the disruptions to everyday social order showed that social order was maintained in the minutiae of everyday interaction. Such interaction was marked by common assumptions, and contextual features (e.g. about the nature of a social episode such as 'watching television') were mobilized ad hoc to sustain the sequence of events (e.g. that make up 'watching television'). The participants in – or the subjects of – the breaching experiments that disrupted everyday episodes assumed that things would revert to normality and that the disruption was a deeply moral matter that demanded explanations from the breachers (Heritage 1984).

Part and parcel of such everyday interactions is that coparticipants trust one another to fill in the gaps that make up such interactions – the unarticulated tacit common assumptions. If those assumptions are questioned, as in breaching experiments, people see this as highly disruptive and dangerous. This implies that all action is accountable. This applies no less in specialist social domains such as in the scientific laboratory (e.g. Lynch 1985, 1993). However, as Heritage (1984) forcefully notes, for ethnomethodology there are no extraneous rules that attach to a context and tell us what actions are appropriate. Action is seen to be in-process, and thus unfolds and creates its own context: it gathers up past actions into current ones and defines or redefines them according to the current situation or interaction. 'Rules' or conventions or expectations about the right way of behaving are mobilized in accounting for action. In a sense, as everyday actors we ask ourselves, what current and new actions do past ones allow that can be accounted for, warranted and understood convincingly.

Like ethnomethodology, conversation analysis (CA) assumes that both the production of conduct and the interpretation of conduct are accountable and part of a common everyday set of methods and procedures. At the most basic level, CA is concerned with the ways that the current production of conversation is a local, here-and-now definition of the situation and that subsequent talk will be orientated towards this. At an elementary level this means that some current 'turn' in talk implies or projects a relevant next activity or range of activities – that is, bits of talk. This has been called 'sequential implicativeness' (see Schlegoff and Sacks 1973) and finds its simplest expression in pairs of actions such as ritualized greetings (hello – hello, goodbye – goodbye), question-answers (request – grant/rejection) and invitation (acceptance/rejection).

These 'adjacency pairs' can be said to adhere to the following rule: 'given the recognizable production of a first pair part, on its first possible completion its speaker should stop and a next speaker should start and produce a second pair part from the pair type the first pair part is recognizably a member of' (Schlegoff and Sacks 1973, p. 296). As Heritage (1984) remarks, this looks like a very complicated way to say that we exchange greetings or that questions are answered. However, there are a number of important insights embedded within this formulation. There is no claim that a greeting invariably follows a greeting – this is not a mechanical rule. Rather, a greeter acts on the presumption that a greeting always proposes that a return greeting is due next. This means that, when making a greeting or asking a question, we assume that the appropriate responses are due. Such pairs are thus considered normative; in producing a greeting, one is proposing that a second speaker return the greeting immediately the initial greeting is completed – that this is accountably due, in the sense that if there is a deviation something needs to be done about it. Conversation analysts have traced in many different ways the means by which apparent deviations from such norms are no such thing (there are circuitous routes by which the adjacency pair is re-invoked or 'repaired').

Now, these are seen to be basic taken-for-granted dimensions of everyday life. There are various other attempts at getting at such fundamental features, perhaps most famously exemplified in the work of Goffman, who attempted, but failed, to abstract a basic schema (such as his dramaturgical model) through which to analyse all everyday settings (see, for example, Goffman 1959, 1961, 1964; Manning 1992). Crucial here is that these attempts to grasp the basic *forms* of social interaction can be regarded as 'timeless' and 'placeless': they address the fundamental features of human social interaction and thus are common to all groups at all times. As such, there is nothing that is not 'everyday life'. Stephen Crook (1998) frames it thus: 'a formal "taken-for-grantedness" can be shown to be a presuppositional dimension of *any* region of social life ... [as such] everyday life ... expands to become equivalent to social life as such' (p. 528, original emphasis).

These formalistic accounts of 'everyday life' have however been subjected to much criticism. The usual complaint is that they are 'descriptive' (e.g. Gouldner 1975; Gardiner 2000). They do not attempt to understand the particular substantive causes and historical circumstances of particular substantive versions of everyday life, nor to mount a 'critique' of these versions – that is, to expose their problems and point to, or at least acknowledge the possibility of, versions of everyday life that are 'better' for its inhabitants. However, Crook (1998) argues that separating the formalists and the 'substantivists' (or critics) is an unbreachable divide between form and substance. We might add that this unbridgeability can be expressed as an infinite regress. On the one hand, a 'formalist' argument would regard critics' accounts as exemplifying the formalist argument (as I have done above with regard to the everyday warranting of heroic writing on the everyday). On the other hand, the critical view would regard the formalist account of the social world, and of its account of warranting in critical academic accounts, as indicative of the historical and critical blindness of the formalist perspective. This is an impasse that cannot be broken unless one can accept the mixing up of form and substance – or the 'minotaur', as Crook envisages it. However, the minotaur is not unassailable (see Sorensen 2004). By way of a foretaste, we can note that the minotaur is not necessarily the impasse it would appear, especially if we begin to see in the processuality of everyday life the taken-for-grantedness of relations of power, and their sometime contestation. That is to say, if everyday life is characterized by ordering and disordering, the performativity of 'efforts'[3] at ordering and disordering affect other like 'efforts'. Crucially, such everyday efforts include critique of others. Put simply, critique can be done by 'ordinary people' as well as scholars. In Crook's terms, substance is part of the form of everyday life. We shall pick up on this issue in Chapter 8.

The everyday as elusive, fragmented and heterogeneous

For the formalists, the everyday world is necessarily everywhere. One empirically studies it in its various manifestations, and making broader theoretical or substantive claims is held in check, not least because such claims would need to be subjected to the same formalistic analysis. In contrast, for the substantivists it poses a major problem of access. For such recent commentators as Highmore (2002), situated in a crooked lineage that includes the surrealists, the situationists, Simmel, Benjamin, Lefebvre and De Certeau, while the everyday is characterized by the ordinary, the boring, the obvious it also has at its heart a strangeness. As Highmore (2002, p. 16) puts it, 'The everyday offers itself up as a problem, a contradiction, a paradox: both ordinary and extraordinary, self-evident and opaque, known and unknown, obvious and enigmatic.' Irruptions of the unconscious, the phantasmagoria of commodities, the displayed exoticism of 'others', the sheer volume of sensation – these make the everyday a terrain of the marvellous. This quality of marvellousness, as Gardiner (2000) notes, is something that the formalistic versions of the everyday fail to attend to.

While Highmore makes no claims to explicating a coherent tradition, he does point to a series of questions that arise if one accepts this vision of the everyday. Crucial here is the issue of how to 'capture' the everyday – that is, how to represent it in a way that does not reduce its richness and contradictoriness. For Highmore, what is needed are something like the techniques of the avant-garde, especially that of montage, which can register the cacophony that comprises everyday life in late modernity. This problem of representation is redoubled when the complexities of what is to count as evidence of the everyday are addressed. After all, the usual forms of textual or visual data (e.g. photos, observations, interviews, letters, media materials) can proliferate beyond any possibility of assimilation – a problem that, as

Highmore shows, dogged the enterprise of Mass Observation that swung between, as much as encompassed, empiricism and 'impressionism'. How does one deal with this profusion of data? This is still further complicated by the fact that heterogeneous data drawn through the other senses – sounds, smells, touch – can also inform the study of the everyday. We might suggest that techniques need, for want of a better term, to engage in something like 'capturing' the everyday. Along with Highmore, we can problematize the idea of 'representing' with its connotation of accuracy – there is too much complexity, fragmentation, elusiveness and heterogeneity to stake such empirical claims. 'Capturing', by comparison, denotes an active, creative and continuous process of chasing, stalking and grappling, while at the same time connoting its obverse, 'escaping'. Highmore detects these struggles to come to grips with everyday life in various theorists of the everyday. For example, in the poetics of De Certeau and his colleagues (De Certeau, Giard and Matol 1998), there is an attempt to capture everyday life through complex weaving accounts that allow the polyphony of voices and activities in everyday life to be expressed, even where these are 'tactical' – ruses, secret, playful, disguised. In all this, the analyst's voice does not take precedence; it is singularized, one among many. A different sort of complexity is present for Lefebvre's version of the everyday: here, even the most ordinary and trivial of activities, such as buying sugar, contains within it a multitude of causes and reasons that can serve as a summation of a society and a history (an epoch, even). To get at these layers, the apparently 'simple' everyday event needs to be defamiliarized through, as Highmore notes, a 'plurality of approaches, a range of attentions that place [the everyday] within a framework of critical interdisciplinarity' (2002, p. 143).

As with Highmore, something like a process of 'capturing' is the first task of studying everyday life. Indeed, it needs to precede critique not least because the very process of 'capturing' might reveal complex moments of 'critique' *within* everyday life (see below).

The everyday as bad: repetition, discipline, alienation

It is commonplace among writers on everyday life to point to the negative dimensions of everyday life. There are several ways in which this is formulated. For Lefebvre (1947), for example, everyday life in modernity is characterized by alienation in which human potential (associated with the figure of 'total man' (*sic*)) is dissipated in a society defined by rigid roles, and by commodification in which exchange value takes priority over the use value of objects. In a later work, Lefebvre's (1984) version of modernity becomes still darker: it has become 'the bureaucratic society of controlled consumption', a phrase that Lefebvre (1947, p. 60) believes encompasses the following features: 'the limits set to (society's) rationality (bureaucratic), the object of its organization (consumption instead of production) and the level on which it operates and on which it is based: everyday life'. Moreover, such a society tends towards what he calls 'a terrorist society' in which

> bureaucracy binds the 'individual' hand and foot by total exploita-
> tion ... by turning (people) into bureaucrats (thus training them for
> the bureaucratic administration of their own daily lives) and ration-
> alizes 'private' life according to its own standards ... persuasion turns
> into compulsion – which is an exact definition of terrorism. The only
> perspective open to the (inner) gaze are the avenues of make-believe,
> the only possibility of (illusively) adapting to circumstances appears
> to be ... through consuming the symbols of violence and eroticism
> made available to the public. (pp. 159–160).

Now, Lefebvre (1984) notes that everyday life cannot be wholly bureau-cratized: 'residual and irreducible, it [everyday life] eludes all attempts at in-stitutionalization ... [it is] ... the time of desire: extinction and rebirth' (p. 182). There are a number of points to be clarified here. First, for Lefebvre, premodern everyday life is a domain of natural rhythm and repetition that is related to labour (the seasons, the diurnal), life and death. This contrasts with the complex dynamics of contemporary everyday life, which has become divorced from these rhythms (and the 'needs' associated with them). One might almost read Lefebvre's remark concerning the process of adaptation to bureaucratization (the consumption of symbols of violence and eroticism – what the situationists might call 'spectacle', see Debord 1983) – as a distorted echo of those rhythms and needs. Second, at the same time, the everyday manages to elude such bureaucratization by virtue of being a 'time of desire'. On this score, the premodern is still present in the late modern (this is also the case for certain social forms such as the 'festival', whose relation to 'spectacle' is problematic, to say the least).

So, for Lefebvre, the modern everyday seems to be an ambiguous mixture – largely alienated, but stirred by the echoes of the premodern. On this score, there are parallels to De Certeau's (1984) account where deep-seated memo-ries resource 'tactics' (see below). If Lefebvre's analysis focused on bureau-cratization (and institutionalization), De Certeau's perspective on everyday life was driven in part by a reaction against what he saw as Foucault's (1979) reduction of everyday life through a single procedure – panopticism – in which, famously (and crudely), people were subjected to a constant surveil-lance that disciplined them into 'docile bodies'. For De Certeau this was a *strategic* process that depended on a demarcated spatiality within which dis-cipline can be practised. His task was to excavate the tactics by which such discipline is 'fooled' – tactics that were resourced by, for example, family traditions.

The ambiguity that Lefebvre detected in modernity was not uniformly distributed: it was the preserve of only certain inhabitants of modernity. Women most especially seem to be excluded from the capacity to elude, or critique, the exigencies of the everyday. Featherstone (1992, p. 161), in characterizing the broad parameters of everyday life, notes that as a 'sphere of reproduction and maintenance' everyday life is often seen to be the domain of women. In Lefebvre,

> [e]veryday life weighs heaviest on women ... they are the subject of everyday life and its victims or objects and substitutes (beauty, femininity, fashion, etc.) and it is at their cost that substitutes thrive ... Because of their ambiguous position in everyday life ... they are incapable of understanding it. (1984, p. 73).

Both Felski (1999–2000) and Bennett (2004) cite this passage but to somewhat different purposes.

Felski seeks to reverse the valency of repetition in everyday life. Thus, the equation between women, everyday life, repetition and, in the end, collusion with capitalism is a modern artefact. Repetition and routine entail 'a much more complex blend of the social and the psychic' through which we (and she means people in general) 'become who we are' (1999–2000, p. 21). Further, Lefebvre's tendency to link 'repetition with domination and innovation with agency and resistance' (1984, p. 21) is spurious, not least as repetition and innovation (and circular and linear time, Felski adds) have a complex relation to the everyday. Sometimes, routine is an invaluable means of staving off unwanted innovation (not least when we turn to technoscience). Drawing on a number of other commentators, notably Heller (1984) and Silverstone (1994), Felski advocates a view of the 'everyday-in-itself': an ontological bedrock of habit, routine and repetition that we cannot ever hope to escape but that is an inherent aspect of being-in-the-world. Here, we see resonances with the formalist conceptualizations of everyday life, though Felski also wants to recognize that everyday life is a domain in which human potential is constrained and thus subject to critique (not least feminist, as in the work of Dorothy Smith, 1988). Here, Felski follows a similar path to Heller (1984, p. 119), who develops the notion of objectivation-for-itself, which reflects conscious human intention acting upon the ontologically prior processes of objectivation-in-itself – that is, the routine, repetitive and habitual in the everyday.

Bennett (2004) by comparison draws on Lefebvre's quote to problematize the very idea that there can be an agent of change that challenges repetition and routine in the everyday. For him, such a figure rests on a wrong-headed notion of everyday life as an object of critique. While he shares with Felski a more complex understanding of everyday life (not least in terms of its temporal structures), he suspects that she too falls, albeit in a roundabout way, into the trap of critique. Objectivation-for-itself, rather than an add-on, might be seen as intrinsic to the domain of everyday life, which can be re-conceptualized as altogether more fragmented, disorderly and reflexive. At this point, we can turn to the use of critique, our last theme in our survey of accounts of everyday life.

The everyday as good: critique, promise and resistance

'Critique', as Gardiner (2000, p. 6) puts it, aims to 'problematize everyday life, to expose its contradictions and tease out its hidden potentialities, and to raise our understanding of the prosaic to the level of critical knowledge'. Now, for Habermas (1987; see Featherstone, 1992; Crook 1998), everyday life – as a communicative-theoretic lifeworld in which shared assumptions, not least about the forms of interaction, underpin cultural reproduction – comprises critique in its own right – critique of those systems (economic, bureaucratic) that would seek to colonize it. For Maffesoli (1989, 1996), everyday life is a Dionysian domain of polydimensional sensuality. As such, it can be turned to a critique of those modernist processes of rationality and power. However, as Crook (1998) points out, Maffesoli is more interested in the celebration of this 'pagan' component of the everyday than launching a critique.

By contrast, rather than everyday life per se resourcing critique, it is typically the object of critique. To be sure in its interstices can be found forms of 'practical critique', but generally speaking it is regarded by critics as a domain of passivity, habit, alienation, and so on. It is instructive to contrast the humanist Marxist version of critique that we find in Lefebvre with the post-structuralist perspective of De Certeau. As we have noted, for Lefebvre everyday life cannot be wholly bureaucratized; it can escape through 'moments of presence', as Shields (1999) explicates. These are transcendent episodes, which (like Dada events) puncture the taken-for-granted, that though the ephemeral nevertheless provide an insight into the totality of society (which for Lefebvre was open and flexible) and a glimpse of the 'total man' (*sic*) in which human potential can be fully realized. Such moments emerge in the process (or dialectic) of everyday life, though they are detached from the temporal flow of everyday life. Further, such moments can echo the pre-modern festival (whose Dionysian ecstasies can be regarded as a more or less unexceptional feature of everyday life). This analysis was, Shields notes, criticized by the situationists, who saw it as overly abstract: for situationists such as Debord (1983) and Vaneigem (1983; see also Knabb 1981 for details of situationist techniques), such 'moments' were a practical matter of re-envisioning the social world, even as such interventions were liable to be recuperated by the state and capital as spectacle.

Against Lefebvre's 'immanent utopian' perspective, De Certeau's analysis of the way in which the everyday incorporates the possibility of different forms of life is altogether less dramatic. As we mentioned above, De Certeau was interested in the way that cohabiting with the disciplinary strategies of everyday life were resisting tactics. The 'tactic' is a practice of the weak that cannot quite be contained by the spatialities in which surveillance takes place. It is fundamentally temporal, emergent, 'always on the watch for opportunities that must be seized "on the wing"' (De Certeau 1984, p. xix). Further, any gains are always temporary and must always be made anew. Such ways of operating De Certeau likens to a number of practices: ruses, tricks, hunter's cunning, joyful discoveries, and so on. One finds such tactics woven into the fabric of everyday practices such as reading, cooking, shopping, idle walking where people ongoingly make up their trajectories as they move through a text, a kitchen, a neighbourhood, a city. Taking the example of cooking, if the recipe is a strategy that specifies a set of actions (and as such disciplines the space of the kitchen and the cook's body), as De Certeau *et al.* (1998) show, in 'doing-cooking ... tactics are invented, trajectories carved out, and ways of operating individualized'. But lest we think of this as merely an expression of the individualization that reflects the sort of mobilities associated with late modernity (see Giddens 1992; Beck and Beck-Gernsheim 1995, 2001), De Certeau *et al.* make it clear that such tactics are 'rooted in family or regional oral tradition' (1998, p. 201). However, in light of contemporary theorizations of consumer society, the albeit muted optimism embodied in De Certeau's notion of tactics should arguably be treated with considerable circumspection. In such a context, these ruses and their embeddedness in family or regional traditions are now ripe for commodification as 'lifestyle choices': governance is no longer a matter of surveillance, rather self-government is mediated through forms of consumption (Boyne 2000). That is to say, instead of the governmental production of docile bodies, the modern western world is replete with dispersed and disparate techniques of governmentality through which people actively 'make themselves' (Rose 1996, 1999; Dean 1999). We shall explore this point further in Chapter 4 when we consider aspects of the micropolitics of modern-day surveillance.

I have remarked that the differences between Lefebvre and De Certeau are notable. Lefebvre situates the 'other' of repetitive, alienated everyday life in the irruption of 'moments of presence' that point to a 'better social world' (however open-textured and hard to grasp it might be). De Certeau locates the 'other' of a disciplined, surveilled everyday life in minuscule tactical ways of operating. In the former there is a distinct utopianism, in the latter a muted celebration. However, both share the rhetorical or narrative structure in which a sorry 'normal' everyday is enlivened by an-other everyday life signalled respectively, by moments and tactics. As Bennett (2004) notes, linked to such general arguments are judgements of what sort of actor might realize or manifest this 'other'. For Lefebvre, as Bennett shows, such agents might be students, or the working class, or youth. For De Certeau, we might say that it is anyone whose roots in family or region remain intact (something that, arguably, is less in evidence now than at the time of De Certeau's team's studies). In both cases, a 'default' everyday is assumed in which the norm is repetition or discipline, habit or order. By contrast, as Bennett (2004; also Michael 2000a) argues, the everyday can be conceptualized as an altogether more complex domain: discipline generates disorder as well as order; repetition leads to invention as well as alienation.[4] Moreover, as I have commented above, intrinsic to the modern everyday are the regular intrusions of the novel and the exotic, and the promise of more or less fundamental changes to 'the modern way of life'. As I shall detail in due course, such intrusions and promises derive in substantial measure from the operations and circulations of technoscience. In sum, prospective disorder is part of present ordering. This all complicates the everyday to the extent that the 'bi-polar' logic (as Bennett puts it) that typifies the critique of everyday life begins to seem less than helpful.

In this section, I have tried to come to grips with some of the key themes that emerge in the literature on everyday life. To reiterate, I make no claims to comprehensiveness – this is a 'survey' rather than a review. As will be obvious, the present perspective on everyday life takes a rather more microsocial line that insists on treating the processes by which the everyday is ordered very seriously indeed. Further, as should also be clear, it is assumed that the everyday is a domain that is routinely disordered – or, rather, in which dis-ordering processes are also at play, some of which might indeed illuminate other forms of social life.

Grasping technoscience

At its simplest, the term 'technoscience' denotes the interweaving of science and technology. Accordingly, science would serve as the empirical and theoretical basis upon which technology is innovated or developed. A discovery in theoretical physics or in some empirical branch of physics might thus be seen as responsible for the invention of innumerable modern technologies. For instance, without Einstein's famous formula $e=mc^2$ it is often claimed that such technologies as the television would be impossible to invent. Conversely, science is nowadays impossible to imagine without the input from an enormous range of specialized technologies, from humble microscopes to particle accelerators. Indeed, much science is about the production of technologies for use in the laboratory. For example, Rheinberger (1997) traces the way that the vagueness and unpredictability of 'epistemic things' – physical structures, chemical reactions or biological functions – are transformed into the discreteness and predictability of technical objects that, rather than being the topics of study, become components in the experimental systems that investigate other 'epistemic things'. That is to say, where once epistemic things 'present[ed] themselves in a characteristic irreducible vagueness ... [because they] ... embody what one does not yet know' (p. 28), once known (once they are no longer open, or immanent), they can become technological artefacts that serve in the doing of subsequent experiments. Thus X-rays move from being 'vague' to being 'tools'. Ironically, the emergence of such technological tools is often a matter of trial and error in which theory features only tangentially.

This relation between science and technology is complicated somewhat when we go on to consider how science is mobilized to render judgements about technologies and their relative strengths and weaknesses. This complexity becomes most evident in relation to those technologies that have allegedly[5] gone more or less disastrously wrong such as Challenger, Bhopal, Love Canal (the list is considerably more extensive, especially when technological developments – for example, genetically modified organisms or new airport runways – are also being assessed in relation to the risks they might pose in the future). Here, science is often rhetorically abstracted as an arbiter over the 'truth' of a technology (whether it works properly or not). Yet science is always a technologically embedded process, and many of the arguments that accompany such assessments concern the credibility of the techniques and technologies that enable scientific judgements to be made. For instance, Collins (1988) traces how the UK's (then) Central Electricity Generating Board attempted to demonstrate the 'safety' of its nuclear fuel flasks in what was represented as an open experiment that provided a genuine test of the flasks. As Collins argues, this was in fact a 'display' of postclosure knowledge in which the uncertainties that would be expected in an experiment had been systematically controlled. As such, this was not an experiment but, Collins insists, a display of virtuosity. The public was denied this sense of the artificiality of the display. Collins' suggestion is that the public should be represented in the assessment of such 'scientific tests of technologies' by counter-experts (e.g. Greenpeace) who can unpick and scrutinize the parameters and contingencies involved in such a display-test. Here, then, technoscience – as the indissoluble interweaving of science and technology – is primarily the domain of experts.

However, commentators have been highly critical of this and related suggestions about the assessment of technology, and this demarcation of a distinctive domain of technoscience (e.g. Wynne 2003). This is an issue that will be elaborated further in Chapter 4. Suffice it to note for the moment that, in light of contemporary shifts in the contexts of technoscience, not only are the differences between science and technology increasingly difficult to sustain, but so too are those between technoscience and, for want a better term, society. Felt (2003, p. 32) puts it thus: 'if one assumes that the meaning of "technoscience" is not fixed in time and is subject to negotiations and if there is no obvious clear-cut societal set of demarcation criteria to distinguish science from non-science, then the public representations constructed in the course of science–public interactions come to play an important role'. As recent commentators have noted (e.g. Nowotny, Scott and Gibbons 2001; Irwin and Michael 2003), for a whole range of reasons, technoscience and society can be said to be in the process of convergence. This clearly has implications for the way we need to formulate 'technoscience and everyday life'. For example, even those technologies that are so mundane as to be more or less invisible (e.g. velcro or air conditioning) fall under the rubric of technoscience: on the one hand they were once, as new innovations, overt products of technoscience; on the other, they are always liable to become part of technoscience once more – say, when they go wrong and need to be reassessed for risks, or when they become objects of improvement, or artefacts that signify a way of doing things that now needs to be 'overcome' (Michael 1997).

This erosion of the specialist domain of technoscience and its boundaries with society, is also evidenced, though in a different way, in other formulations of 'technoscience'. Bruno Latour is generally credited with introducing a more radical use of the term 'technoscience'. For Latour, to study science, and in particular the laboratory with its capacity to generate knowledge and change the world, is to engage with technoscience, which he characterizes in terms of 'all the elements tied to the scientific contents no matter how dirty, unexpected or foreign they may seem' (Latour 1987a, p. 174). That is to say, to study science in the making, it is necessary to follow scientists and engineers as they attempt to order and align people, equipment, funds, representations, texts, entities such as microbes, scallops, electrons, and so on and so forth (what Law (1987), has called 'heterogeneous engineering'). Part and parcel of this analytic is an agnosticism as to what counts as science and what as society (Callon 1986a, 1986b). And just as scientists align disparate entities in the production of knowledge and the exercise of influence, so too do others – politicians, regulators, venture capitalists, bureaucrats.[6] Clearly, this perspective (sometimes called actor–network theory) is orientated towards the microprocesses (for example, forms of persuasion) by which certain actors (technoscientists) marshal and order others (which might be either human or nonhuman). In all this, it is assumed that the social world emerges out of these circulations.

One of the products of technoscience is the accelerated mixing up of the human and nonhuman. Hybrids are everywhere, according to Latour (1993), and while we might try to ignore them they increasingly impinge upon our everyday lives. Here, then, technoscience affects everyday life in a multitude of ways – laypeople, as actual and projected 'users' of technoscience, are 'recruited' into the circulations and orderings of technoscience. That is to say, the microprocesses of everyday life (which, as we shall see, also entail macro views of the social world) are crucially mediated by the doings of technoscience. At the same time, everyday life is coming to be overtly populated by hybrids that, some claim, make any notion of a pure human or a pure society increasingly difficult to sustain (e.g. Michael 2000a; Waldby 2000).

Donna Haraway (1997) has taken up and expanded the term technoscience beyond a conception of circulation of artefacts (texts and materials) which at the micro level serve to make (or unmake) the social (or, rather, heterogeneous) world. In a manner that echoes the substantivist critiques of the formalist accounts of everyday life outlined above, she has criticized Latour (and the sociology of scientific knowledge more generally) for the lack of critique (in fact, his is a principled refusal of critique, one that echoes the position of the philosopher Michel Serres – see Latour 1987b). On this score, Haraway situates herself not only in the tradition of science and technology studies, but also in those of feminism and cultural studies. For Haraway, technoscience is a late modern phenomenon intimately tied to what she calls the New World Order Inc (of transnational capitalism) that not only collapses the distinction between technology and science, but also between human and nonhuman, nature and society, factual and artefactual, subjects and objects. Through the complex narratives and practices of technoscience emerge complex hybrid figures such as the cyborg (Haraway 1991) and the oncomouse, whose ontological and political status is highly ambiguous. Like Latour, Haraway wants us to rid ourselves of our fear of the mixed – the impure – and acknowledge these hybrid products of technoscience. For Haraway, this is a fraught process (see also Latour's more recent formulations, 2004a). The cyborg, for example, is a female figure that is both 'real' in that it marks the embroilment of women in technoscientific circulation, but also 'fictional' – realized through the tropes that come from science fiction as much as from science. The cyborg's political ambiguity straddles, as Prins (1995) excavates, 'the anti-humanist Nietzschean ethic of resistance and self-affirmation' and 'a socialist-feminist ethic of solidarity, a Christian feeling for a suffering humanity'(p. 361). The oncomouse (a mouse genetically altered to grow cancers) induces similar ambivalence. As Myerson (2000) notes, for Haraway the oncomouse is mixed in a plethora of ways – biology and culture, commercial and academic, an index of creativity and a mark of exploitation. Such beings – both the object and tool of scientific study – are at once a product of nature's creativity that is teased out by human technoscientific creativity, and objects through which we come to understand nature. These beings are not only present in everyday life as spectacles (the mediated outrage directed at the oncomouse or the earmouse), but also impact in complex and contradictory ways on everyday conceptions of the body, nature, health, identity, and so on. As Myerson frames it, 'we can acknowledge our kinship with these new possibilities, either as victims or as heroes' (2000, p. 73). Here, then, technoscience produces complex semiotic-material entities that enable it to make and understand nature; yet to explore such figures as the onco-mouse is to interrogate the ways that the everyday is embroiled in techno-science, both as resource and critique. If technoscience teases everyday categories, it is also subjected to a more or less keen scrutiny that touches on matters of trust, disgust, hope, ethics, economics, regulation, to name but a few.

In light of the foregoing discussion, it is worth attempting, no doubt foolishly and feebly, a rough formulation of technoscience in relation to the present concern with everyday life. We have seen how technoscience is en-tailed in the erosion of a series of everyday categories – between science and technology, between science and public, between nature and society, be-tween knowledge and politics. Technoscience can be said to be the circulation of artefacts (these might be textual or technological, but they are always both semiotic and material) that are characterized by, or related to, some form of 'technical knowledge'. Of course, such knowledge is never purely technical: on the one hand, it entails elements and vestiges of common sense knowl-edge and know-how and, on the other, it is resourced by such genres as science fiction. Moreover, such artefacts pertain to the understanding of, or intervention in, what is taken to be the natural world (which includes en-vironments, human and animal bodies, chemical and physical processes, even as these themselves have emerged through prior technoscientific understandings and interventions). As noted, such circulations contain the trace of nontechnical knowledges, yet arcane knowledge is still a defining characteristic, and it is a chore for non-technically trained folk to come to grips with it. Having noted this, technoscience also comprises circulations of its own image: technoscientific artefacts – from the most 'mundane' to the most 'exotic' – circulate alongside accounts of the relations between tech-noscience and non-technoscience. These might take the form of instructions (how to use artefacts 'properly'), of claims to a privileged capacity to assess whether an artefact works or is risky, or of broader demarcations of scientific expertise and public ignorance. In all this, technoscience as a circulation necessarily ties together what have previously been seen as separate – the human and nonhuman, the natural and the social; or rather it is con-stitutively productive of hybrids whose ontological and political status is ambiguous and multiple.[7]

This tentative formulation does not address itself to the specificities of what of technoscience is encountered in everyday life. To compose an ex-haustive list seems to me to be a very difficult, if not an impossible task – not least because the items on such a list would not be discrete and the manner of their blurring would be manifold. Instead, we can pragmatically identify a number of key sites (or points in circulation) at which technoscience (espe-cially as it relates to everyday life) can be 'located'.

- As we saw above, a central site for technoscience is the laboratory in which scientific knowledge or technological innovation is 'produced'. The scare quotes are in place to remind us that what is to count as knowledge or invention is something very much dependent on successful circulation, both prior to the laboratory and afterwards (e.g. Brannigan 1981). Nevertheless, the laboratory remains a key site that, as Latour (1987a) argues, has the extraordinary quality of gathering together a hugely heterogeneous range of materials – money, technology, humans, texts, bodies, chemicals, and so on – and combining them in ways that generate what can be seen to be new knowledge or novel techniques or innovative technology.
- The everyday sense of technoscience is in considerable part derived from the media representation (or the announcement) of 'spectacular' developments or breakthroughs (these might include the invention of a new biomedical technique or technological artefact, the discovery of a new natural fact, the presentation of new images of nature). Clearly, as I shall illustrate, such representations and the claims they contain are highly contestable, not least as what is to count as invention or discovery is highly pliable.
- Technoscience is mundanely manifested in the practical and often unnoticed technologies (and expertises) that cohabit the everyday with us: refrigerators, computers, cookers, radios, televisions, washing machines, beds, shoes, tables, armchairs, clothes, mirrors, envelopes, pens, paper, light bulbs, cars, trains, knives, zips, velcro, ring-pulls ... here the list tumbles out of control. I include these as parts of technoscience because they were once, in one way or another, before they became more or less invisible technologies, inventions that had to be developed and refined and made 'workable' – both technically and culturally. Moreover, many still undergo a process of technoscientific refinement and are the subject, every now and again, of reinvention (e.g. ceramic knives, nanotechnological cloth, MP3 players, and so on). Further, such technologies are not simply material artefacts – they incorporate a cultural baggage that 'scripts' how they might be used; as such, we see how the assumptions of technologists (and designers) inflect with everyday exigencies.
- Behind these more or less invisible technologies are the more or less invisible technoscientific processes of standard setting and maintenance in which what is to count as a safe or efficient or effective or reliable or uniform artefact (or service) is worked out. Examples here include comfortable indoor temperature, or a safe drug (or useful drug trial), or a volt, or air or water quality. While such processes normally proceed behind the scenes, every so often they erupt onto

centre stage, not least when things go catastrophically wrong and some sort of inquiry is required.

- The more or less overt attempts by 'technoscience' to make particular forms of responsive publics, most especially nowadays to make the scientific or technological or technoscientific citizen. Here, publics are constructed in various ways in relation to scientific expertise: increasingly they are invited to deliberate over, or participate in, the making of scientific decisions or policies. Of course, the obverse also needs to be considered, namely the way that certain publics actively enter, more or less forcibly, into the technoscientific decision-making process. In these last two examples of technoscience, the process of technoscientific circulation itself becomes the object of reflection and contestation.

Needless to say, these manifestations of technoscience are far from distinct (for example, standards are routinely tested in laboratories; breakthrough announcements in the media are often associated with surveys of public opinion) and not exhaustive (as promised in Chapter 1, missing is any mention of the role of 'economics' in the process of circulation, or an account of how technoscience in modern western societies is related to its presence in developing countries). Despite its shortcomings, this list of 'sites' should give some hint of the key means by which technoscience folds into and circulates through the everyday, and can serve as a heuristic by which to delineate such folds and trace such circulations.

Some related literatures

Over and above the literatures on everyday life and technoscience, there are several others that straddle these areas. Again, no claim to comprehensiveness is staked, but mention should be made of some of these general perspectives if only to signal that the standpoint developed here by no means exhausts the range of ways in which 'technoscience and everyday life' might be addressed. I draw attention to the following areas of study (which, to be sure, are not discrete): 'technoculture', 'material culture', 'technological domestication' and 'scientific citizenship studies'.

The first area covers a diverse range of approaches that has roots both in science studies and, crucially, in cultural studies. Menser and Aronowitz (1996) note in the opening chapter to the volume *Technoscience and Cyberculture* (Aronowitz, Martinsons and Menser 1996):

American culture is technoculture, from boardroom to bedroom. This is not to say there is just one American culture; there are many, yet each is a technoculture. Truckers and cyberpunks, rap musicians and concert pianists, even hippies and Amish all employ technologies in such a way that their cultural activity is not intelligibly separate from the utilization of these technologies. (1996, p. 8)

Obviously this would apply to any culture. The focus here is on the complex relation between technology (and science) and culture, and even in this one volume we find such topics as the ways in which particular technocultural artefacts (e.g. an environmental policy statement, TV representations of the smart bomb) serve in, for example, the reproduction of the military-industrial complex, or in the deskilling of workers, or in the state production of cultural space. Donna Haraway is a major figure in this tradition (see especially Penley and Ross 1991) and, as already noted, her work informs, in one way or another, much of what is to follow.

In addition, one can also point to those studies that consider the relations between science, technology and such cultural productions as literature and film – for example, the way that certain motifs in science and technology are played out through fiction (e.g. Graham 2002). Of course, these fictional accounts, in turn, feed back into arguments about technoscience itself (the Frankenstein motif perhaps being the most obvious – see Mulkay 1997; Turney 2000; Fukuyama 2002 – though none of these writers would see themselves as practitioners of cultural studies). As such, the social meanings through which both technoscientifically novel and commonplace objects such as the 'gene' are understood emerge out of the complex interplay of science, technology and culture (Nelkin and Lindee 1996; Keller 2001). Again, this work informs what follows, especially the discussion of the way that novel technoscience is presented to, and impacts upon, publics.

The second general perspective I draw upon also has roots in cultural studies (though in other disciplines too, notably cultural anthropology). Having located the study of material culture in this way, it is (rather like the study of technoculture), to quote Miller (1998), a somewhat 'un-disciplined' enterprise. Nevertheless, key to this perspective is the focus upon the material object itself in order to trace the specific role it plays in a particular culture. The objects involved here are innumerable – a (not so) random list might include Coca-Cola bottles, microwave ovens, carpets, customized cars, the Sony Walkman, clothing, the dressing table (see Miller 1998, 2001; Dant 1999; Attfield, 2000; Bull 2000). In exploring the cultural role of such things, the concept of 'consumption' has become pivotal, not least because 'consumption' denotes both the realization of certain instrumental functions, but also the production and reproduction of complex and contradictory identities and relations of power (see, for instance, Lury 1996; Slater 1997).

Associated with this perspective is the study of how technologies become more or less integrated into local, mundane settings and routines – that is, how such technologies come to be 'domesticated'. Here, the focus is on how technologies impact upon, and are reformed both materially and semiotically by, people (along with their existing technologies) in the course of everyday life. As Lie and Sorensen (1996) note, users at once shape and are shaped by new technologies as they go about, in the terms of Silverstone, Hirsch and Morley (1992), 'domesticating' them. In the process of such taming, 'users/ consumers make active efforts to shape their lives through creative manipulation of artefacts, symbols and social systems in relation to their practical needs and competencies' (Lie and Sorensen 1996, p. 9). As we would expect, such efforts are highly variable: at times, the introduction of such new technologies is ostensibly seamless in that they fit snugly into existing practices; at others, such technologies apparently aggravate relations of power and necessitate effortful adjustments. We will come across examples of domestication that range from the seamless to the fraught in the course of this book. As with the material cultures perspective, 'consumption' is a key motif: people domesticate such technologies in ways that realize practical and expressive (e.g. concerned with identity, status) functions. However, the way that the introduction of technology affects pre-existing functions and identities – indeed, relations of power – is by no means straightforward. Thus, Lie and Sorensen (1996) insist, in relation to a category like gender, new technology does not have an always predictable impact: both gender and technology re-emerge in the process of domestication such that, in some instances, existing gendered relations are reinforced, in others subverted. The obvious point to note here is that technology's insertion into everyday life is profoundly complex. However, this complexity is underscored when we also take into account the complexities associated with technologies that already occupy the everyday, and the complications wrought by the spectacles of technoscience, not least promises of futures, both good and bad.

The preceding mention of 'consumption' brings us almost neatly to my fourth general perspective, which concerns, broadly speaking, 'scientific citizenship studies'. It is now commonplace to assume that citizenship and consumption interdigitate in complex ways (e.g. Saunders 1993; Gabriel and Lang 1995; Michael 1998a). The choices of what we consume can have political effects, sometimes deliberately so, as in the case of boycotts of the products of certain companies or countries. Conversely, it is often claimed that our politics increasingly reflect the sorts of concerns we have over the objects (and services) we consume, most obviously with regard to the various risks they are believed to pose. Within this context, a context that incorporates a marked scepticism towards liberal democratic politics, there seems to be, in the West at least, a proactive effort to engage the public in the democratic process (Giddens 1998). In relation to technoscience, a number of factors – for example, the general ambivalence towards scientific institutions that writers such as Lyotard (1984) and Beck (1992) detect, or the erosion of exclusive technoscientific expertise in the context of chronic indeterminacy (e.g. Funtowicz and Ravetz 1993; Nowotny, Scott and Gibbons 2001) – have additionally led to publics being canvassed in various ways for their views. Here, the public can be seen to be represented as 'scientific citizens' – or better 'technoscientific citizens' – mobilized by their everyday concerns (about risks, say, though see Tulloch and Lupton 2003), which should feed into the process of technoscientific policy making. However, as will be detailed in due course, these democratizing endeavours are also seen to be problematic, not only practically, but also in terms of the way they come to construct everyday life and its occupants.

These brief forays into the four broad and interlinked areas of 'techno-culture', 'material culture', 'domesticating technology' and 'scientific citizenship studies' underline the complexities that will characterize the analysis of 'technoscience and everyday life'. These areas will be drawn on in an ad hoc manner to illuminate particular issues as they arise in the 'empirical chapters'. Suffice it to say that they inform, albeit tacitly, the next section of this chapter, in which we begin to re-read (aspects of the) literatures on everyday life and technoscience through one another.

Mutual re-readings: some concluding initial thoughts

Technoscience is related to the everyday in its microsocial guise as a process of taken-for-granted ordering, in a number of ways. Needless to say, the norms, expectations, conventions of social interaction are in part enacted with, and mediated through, all manner of mundane technologies. Many of these technoscientific artefacts operate as the facilitative backdrop to the doing of everyday life: door closers and velcro, heating systems and light fittings all render bodies and their senses 'comfortable'. At the same time, these technoscientific artefacts do not simply 'work' – they are not transparent mediators of the everyday; as we saw in the case of velcro, technologies continuously 'go wrong' – or rather, their 'working-ness' rests on myriad minuscule everyday adaptations. In other words, technoscientific stuff not only serves in the ongoing making of the everyday world, it also provides for the (small) moments of creativity (which might remake the everyday). In contrast, certain everyday activities can become revised in more dramatic ways due to the impacts of technoscience. At one extreme, in light of risks identified through technoscience, people can adapt their everyday routines very substantially (e.g. walking in light of the foot and mouth epidemic in the UK, the avoidance of certain artefacts – food, vehicles, pharmaceuticals – deemed dangerous). However, such technoscientific pronouncements can also be 'hopeful' and as such can open up possibilities for positive changes to everyday routines. This sort of adaptation also happens – albeit with less fanfare – when new artefacts enter households and come to be 'domesticated', or rendered familiar in the case of knowledge products. Again, these circulations of technoscientific texts and materials, on one level, need to 'persuade' the occupants of the everyday – but this is a complex process involving heterogeneous re-articulations, including social negotiation in which relations of power (e.g. around gender) can be at once reproduced and reworked. In this process, there is a degree of 'making' people. However, on another level, one can regard such circulations not as 'outside' of everyday life and thus 'entering' into it, but as a constitutive part of it. If the routines of everyday life are already regarded as dynamic, the circulation of technoscientific artefacts (texts, materials) is simply a dimension of the everyday processes of ordering, reordering or disordering.

Attending to the elusiveness of everyday life serves to illuminate technoscience in several ways. Thus, we can highlight the fact that technoscientific artefacts are themselves difficult to pin down. As we shall see further, technoscientific artefacts – though their uses might, in a number of ways, be 'scripted' (Akrich 1992) – can proliferate in their uses and meanings. As such they can be both ordinary and extraordinary, simple and complex. Technoscience as a process of circulation is, as noted above, itself robustly heterogeneous, comprised of the technical to be sure, but also of informal or tacit knowledges, commonplace metaphors, and more or less exotic fictions. If this sheer multiplicity renders technoscience 'elusive', this is redoubled by the chronic uncertainties that nowadays characterize much of technoscience. As I shall have cause to comment, the technoscientific knowledge most relevant to everyday life (e.g. concerning matters of risk) is profoundly uncertain.

Now, when we turn to the difficult themes of everyday life as 'bad' or 'good', along with Bennett, we would wish to dispute the 'bi-polar' modes of thought that have underpinned such valuations. Everyday life is a brew of orderings, reorderings and disorderings that at once empower and disempower people in ways that are not always obvious. In the chapters that follow we will trace a number of ways in which technoscientific circulations enter into these everyday dynamics. In keeping with the general tenor of caution about critique, we shall not rush to identify the ways in which technoscience contributes to, and mediates, 'standard' relations of power (e.g. around gender or class). Rather, as Highmore insists, we first need to trace the complexities inherent in the roles of technoscience in everyday life in order to unravel the ongoing emergence of relations of power, relations that can be unexpected and novel. This gives a different spin to the concern with how the dis/empowerments of everyday life relate to technoscience. To be sure, the circulations of technoscience are hardly uniform, and there are innumerable struggles and contests over access to various technoscientific artefacts (both materials and knowledges). However, these contests often pan out in unusual ways, and the actors that do the struggling can be oddly heterogeneous, changeable and dispersed. To complicate matters still further, technoscience entails the circulation of certain versions of the everyday – versions through which particular and disparate sorts of bodies, citizen, society and the future are performed. These impact on everyday practices in complex ways, not least in potentially advantaging some actors over others, but also in influencing the very 'how' of everyday politics. That is to say, everyday politics is partly concerned with what sort of actors can have what sort of voice and thus 'rightfully' contribute to this or that debate, controversy, argument; in the process of such 'voicing', technoscience serves as a resource, a medium and, indeed, a topic. Finally, we can note that technoscience also resources the way that the everyday 'unfolds': technoscientific futures entail hopes and anxieties that serve in the making of the everyday in the here and now, a making that entails both connection and disconnection, both ordering and disordering. Perhaps, then, we might reframe the 'utopian' moment in everyday life, read through technoscience, as the possibility of a proliferation of dis/orderings. This is something less than Lefebvre's 'total man' (*sic*) and 'moments of presence' (there is no opening onto a totality), but something more than De Certeau's ruses and tricks with their inherent constraints.

The foregoing are a few musings on some of the ways that analyses of, respectively, everyday life and technoscience might inform one another. What I hope emerges is not only a sense of the complex embroilments between everyday life and technoscience, but also a glimpse of the ways in which the partial connections between relevant literatures might illuminate such embroilments. The rest of the book is devoted to continuing this process of illumination.

Notes

1. I should clarify that I use the term 'survey' rather than 'review' because I do not enter into the detail, nor aspire to the objectivity or comprehensiveness, connoted by 'review'. Survey, with its nicely ambivalent hint of distance and situated-ness seems a better term to convey my sense of these literatures as neither coherent nor discrete.

2. Certainly the number of books, let alone articles, on everyday life is growing rapidly. Some recent and forthcoming titles situate everyday life in relation to such social scientific concerns as religion, community, self-identity, consumption, globalization, the body, risk, ethnicity.

3. While the term 'effort' commonly connotes elements of voluntarism and agency, in the present context it is used merely to evoke the heterogeneous dynamics of ordering and disordering in which persons are embroiled. In other words, ordering and disordering are material and semiotic processes of

ensembles or assemblages of elements (or nexus of relations), and these are performative in so far as they might impact upon, or interact with, or merge into, or differentiate from, the ordering and disordering processes of other ensembles or assemblages or nexus. The semiotic aspects of such processes might well include 'mundane critique'.

4.   These points are nowadays familiar from complexity theory (e.g. Urry 2003). Another source for thinking such processes through is the work of Michel Serres (e.g. 1982a, 1982b, 1995b), not least in his commentaries on Lucretius and the clinamen.

5.   I say 'allegedly' because science is also used to stake claims as to the role of human error in the relevant technological disaster.

6.   Though the range of heterogeneous entities available to these actors is rather more circumscribed than that for scientists. It is in fact the extent of such heterogeneity that, according to Latour (1987a, 1990), partly lends science its peculiar potency.

7.   There are interesting intersections with Pickstone's (2000) account in which technoscience 'refers to ways of making knowledge that are also ways of making commodities, or such quasi-commodities as state-produced weapons' (pp. 13–14). He traces the emergence of technoscience to around 1870, which saw the 'development of inventive, intense and self-perpetuating synergies between ... three sets of interests' – namely governments, universities and industrial companies (p. 14). While Pickstone focuses on the complex internal organization of technoscience, the present work is partly concerned with the ways in which technoscience is enabled by everyday life, which it, in part, commodifies. Indeed, in light of Haraway's analysis, we can reverse Pickstone's formulation – technoscience 'refers to ways of making commodities that are also ways of making knowledge, but knowledge not only about the natural world, but also of the cultural'.

3 Technoscientific bodies: making the corporeal in everyday life

Introduction

As I read my Sunday paper, two articles catch my eye. The first, on the front page, bemoans the cuts in hospital cleaning staff, the consequent dirtiness of hospitals and the links to 'the increase of lethal superbug MRSA'. The second article, a sizeable feature on page 4, splashes with the headline article 'Wonder drug that could make heart attacks history'. I read that a pill is being developed that will suppress the production of artery cells that comprise mounds on which cholesterol, for example, can accumulate to produce an obstruction (*Observer*, 9 January 2005). Here we have an example of the extremes of technoscience – both mundane and exotic – as it relates to the body: on the one hand, buckets, mops and disinfectant; on the other, the transfer of the gene for human vascular endothelial growth factor into a virus that can infect artery cells, produce more VEGF, and thus suppress cell production. The everyday features in these reports in the guise of cleanliness and eating habits: what we glimpse in these brief examples is how these normal(ized) activities that care for the body are mediated by technoscience. It is this complex of body, technoscience and everyday life that will be explored in this chapter. In the process, it goes without saying, we venture into the huge and disparate field of the 'sociology of the body' (a list of key latter-day texts might include, for example, Shilling 1993; Synnott 1993; B. Turner 1996), and while no attempt is made to overview this literature, a number of themes central to this body of work are addressed – notably the idea of the openness, or distributedness, of the body, and its enactment.

The body is now a commonplace feature of sociological discussion. After all, everything we do entails the body, despite the projections of certain commentators on the future (see below). As Nettleton and Watson (1998, p. 1) put it, 'Everything we do we do with our bodies – when we think, speak, listen, eat, sleep, walk, relax, work and play we "use" our bodies. Every aspect of our lives is therefore embodied.'

However, embodiment is no simple matter – the body as it is performed in everyday life is realized through its interactions with its environment, an environment populated by the material and cultural products of technoscience. As hinted in the opening paragraph, this 'in-corporation' of technoscience ranges from, at one end, the most automatic and unreflexive bodily comportments, to, at the other, the arguments over the transcendence of the body into something 'post-' or 'trans-' human. In the case of the former, the pervasive presence of technoscience affects such 'basic' bodily reactions as the sense of 'comfort': developments in the technologies of home heating or the particular standardization of settings for air conditioning affect what the body comes to find comfortable (or convenient). In the latter perspective, we find discussion of the body as wetware to be transcended when consciousness is uploaded as software onto the hardware of computer systems, or of the body becoming so manipulable through genetic or technological intervention that the concept of 'human' becomes inadequate to the task of properly capturing the meaning of these new bodies. Between these extremes of deep mundanity and the farther shores of posthuman speculation lie bodies that are embroiled with the more or less everyday technologies of domestic and working life (cars, computers, chairs) and more or less cutting edge technoscientific innovations, most obviously in biomedicine[1] (health communication campaigns through to stem cell therapy, genetic diagnostics, organ replacements).

With this mention of biomedicine we seem to be moving away from 'everyday life' – from a domain of bodily 'normality' to 'abnormality'. Yet, of course, at stake here is what is meant by the body and everyday life. For many, the body-in-everyday-life is characterized by the chronic engagement with biomedicine and its technoscientific products. Even if these latter do not impinge directly on the routines of many people's everyday life, for most of us they are present in the background.[2] To the extent that we now live in a risk society (Beck 1992) – and there are reasons to be circumspect about this model of modern society (e.g. Adam, Beck and Van Loon 2000) – it is commonplace to be aware of risks to the body (whether these risks are to be found in the environment, present in domestic products and arrangements, or derive from biomedicine itself) and the biomedical responses to these. If our newspaper examples stress the biomedical 'performance' of bodies, there are no shortage of examples where the media highlight the bodily risks we encounter in everyday life.

An evocative illustration of this riskiness of the everyday can be found in the UK's Sky One TV programme *So You Think You're Safe*, which looks at the risks posed by, among other things, everyday technologies. At the beonscreen.com website ('the UK's number One site for enabling people to take part in their favourite TV shows'), potential participants in *So You Think You're Safe* are asked such questions as: 'Ever electrocuted yourself mowing the lawn?'; 'Has your office got Sick Building Syndrome?'; 'Ever been strangled by your fairy lights?' In pursuit of the aim to collect examples of mishaps ('the more bizarre the better! – http://www.beonscreen.com/uk/

search/so_you_think_youre_safe.asp, accessed 11 January 2005), we are pre-sented with a picture of the human body under siege: there are risks at every turn that require constant surveillance. To be sure, these programmes aim to draw their appeal from typical 'news values', but the more general point is that this appeal at once reproduces and reflects – that is, performs – a parti-cular version of the body in everyday life that is chronically exposed to risks and the dangers of 'abnormality' and 'pathology'. Of course, for all this normalization of abnormality, this should not detract from the ways in which certain bodies are nevertheless more liable to these 'pathologies' than others.

Here, it would seem that everyday life is linked to a particular model of the body, what we might call the 'body-under-siege'. However, this 'body-under-siege' inflects with other representations through which people enact and articulate their bodies. As we shall see, these representations entail both expert and folk knowledges, though these are not always easy to keep sepa-rate. As noted above, the sense of the body that we possess, and our everyday scrutiny and surveillance of it, are intimately tied to biomedicine: biomedi-cine circulates practices, models, artefacts that crucially shape the everyday apprehension and performance of the body. And yet there are a number of ways in which we must be circumspect about this influence of biomedicine. For instance, we must be wary of reifying biomedicine – or, more mundanely, health provision: as we shall see, there are many variants of doing medicine 'biomedically' and these do not always sit comfortably together. Further, and relatedly, we should resist isolating biomedicine as a separate domain: the models that inform biomedicine themselves have some provenance in aspects of everyday life. Finally, there are ways of practising the body that draw on nonwestern or subordinate western traditions, and such traditions not only resource resistance to biomedical perspectives, but have also begun to inter-digitate with them. In all this, it would seem that the 'body-under-siege' has a, perhaps bewildering, variety of mitigating options available to it.

So far, I have briefly listed some of the ways in which the body comes to be mediated by technoscience (and I include biomedicine in this) in everyday life. However, we should not forget the role of the body in technoscience itself. Despite the apparent 'cerebral' nature of science, the body is crucial in the practices (and discourses) of science (Lawrence and Shapin 1998). Thus scientists must travel from laboratory to laboratory to assimilate the em-bodied or tacit skills necessary to conduct particular experimental procedures (e.g. Collins 1985; Knorr Cetina 1999), and inevitably some scientists are better at some procedures than others. While we cannot follow this up here, we should bear in mind that technoscience itself emerges out of embodied practises, which if better integrated into the self-conception of science could profoundly radicalize it (see Latour 2004b).

This sketch of the multiple ways in which bodies, technoscience and everyday life interconnect is informed by (as evidenced in the use of such terms as 'perform' and 'enact') a conception of the body-in-everyday-life as emergent, relational and distributed, but riven by contrasting comportments and models. In what follows, we shall explore this terrain in considerably more detail. We begin with an examination of the way that the body is 'made' invisibly in its everyday interactions with the mundane objects and techno-logical (or technoscientific) systems. Here we trace some of the assumptions made about bodies, as well as the varieties of comportment that emerge in relation to such technologies. In the process, we take note of some of the 'risks' posed by such technologies to bodies, and contrast these to those that biomedicine 'uncovers' and 'treats'. As a necessary part of this, we go on to explore some of the key representations of the bodies that have informed biomedical interventions, and the way these more or less fold into everyday life. However, we also place these in the context of more 'exotic' rep-resentations in which the body is, in one way or another, transcended, and consider the purchase that such representations have on the popular imagi-nation. Finally, we attempt to pull together these disparate trajectories be-tween technoscience, the body and everyday life by exploring one everyday (or every night) corporeal activity in its complexity (namely, sleeping) in order to posit a 'performative' and 'open/closed' version of the everyday technoscientific body.

Everyday bodily encounters with technology

When we enter into a 'normal' room, a number of conditions need to be met in order for most of us to be able to do things comfortably, that is ordinarily. Lighting levels need to be such that we are able to see; heating levels need to be such that we are able to move with ease (unburdened by too many clothes or too much sweat). Yet as Shove (2003) neatly documents, such conditions are not in any way 'natural', rather they are the outcome of a complex process of co-evolution of devices (e.g. the machinery of air conditioning), systems (e.g. those organizations that have standardized the ideal working tempera-ture at 22°C) and practices (those habits and expectations held by users and consumers). Obviously enough, for some technologies, while they might be standardized, there are local variations in how they are used. For example, people will dress accordingly in order to reach their ideal proximal tem-perature in air-conditioned environments, or use fluorescent bulbs (them-selves technologically highly standardized) in markedly different ways (as in the contrast between Norway and Japan – in the former, fluorescent light is reserved for kitchens and bathrooms, being seen as a cold light; in the latter it is often used in the living room where it is seen to mimic the preferred daylight illumination). Here, bodies (and their senses) are mediated by technologies (and their systems) of temperature and illumination that articulate in complex ways with local customs and expectations, as well as 'globalizing' expert standard setting. In this sense, the body is being constituted in relation to such artefacts: its capacities and capabilities emerge through the routine interactions with technologies. Thus what we experience as 'comfortable' now is different from what we experienced as 'comfortable' a hundred years ago. In these brief examples we are witness to some of the ways in which technoscience interfolds with technologies (conceptualized as 'sociotechnical assemblages' comprised of complex heterogeneous distributions that include humans and nonhumans, social practices and technical artefacts) to impact upon, and more or less 'shape', the body and its comportments.

Latour provides another compelling example of this 'local production of bodies' in his famous discussion of the door closer (or door groom – the device at the top of a door that slowly shuts it, ideally without slamming). Automatically pushing through such doors as we go about our everyday lives, our bodies are being 'made' in the sense that the assumptions built into these technologies require that certain skills and capacities become practised, habitual. Of course, this is not possible for all bodies, as Latour (1992) remarks:

> ... neither my little nephews nor my grandmother could get in unaided because our groom needed the force of an able-bodied person to accumulate enough energy to close the door later ... these doors discriminate against very little and very old persons. (p. 234)

Everyday life is filled with these little interventions that unnoticeably, at least for many of us, discipline – or, better still, enact or perform – our bodies in particular ways. If this suggests a Lefebvrian repetition that 'reduces' the body's capacities (in a trade-off for a certain degree of 'convenience' – e.g. fewer drafts and less slamming), it also points to the way that this closing down is grounded, ironically, in an opening up of the body to a series of broader technoscientific and cultural conditions. As Michael (2000a) has commented, the 'automaticity' and effectivity of such technological artefacts as the door groom (but also an enormous array of other technologies – from zips to ring-pulls, from kettles to the computer mouse) is a complex historical accomplishment. For example, if, as Latour suggests, the door groom is so pervasive because it does the job a 'concierge' once did, the relative cost of labour means that such human door grooms are no longer practicable, unless the aim is to indulge in conspicuous consumption (e.g. the grand hotels).

Moreover, as the examples provided by Shove (2003) hint at, such everyday technologies can be 'usurped'. When they are in some way 'inappropriate', the body can be seen to 'expand' – to draw on more or less deeply ingrained 'ruses', as De Certeau would put it, that recruit other techniques, bodies and technologies in the effort to negotiate what are effectively hindrances. Putting on more or less clothing in air-conditioned environ-ments, or asking passers-by to open a stiffly groomed door – these are the tactics that reinscribe the technology and, perhaps only momentarily, 're-make' the body (or, better still, recollectivize its performance by distributing it across a different array of relations). Of course, those who most notice and respond to these inadequacies in the technology are people with disabilities, who are particularly discriminated against. In the UK, recent legislative moves[3] require the implementation of, where 'reasonable', the redesign of technologies of accessibility (e.g. doors, stairs, lifts, etc.). Such changes, trouble De Certeau's contrast between tactics and surveillance: what were once tactics (e.g. in relation to such sociotechnical discriminations) have been transformed into the remaking (or renormalization) of space.[4]

Now, at risk of stating the obvious, mundane technologies both in their initial introduction into, and circulation within, everyday life make certain 'promises' to, as well as assumptions about, the body. The body, in order 'successfully' to appropriate and use these everyday technologies must con-form in particular ways; at the same time by appropriating and using such technologies it is enabled or enhanced in some manner or other. Inevitably this is a highly complex process as the domestication (e.g. Lie and Sorensen 1996) literature has amply demonstrated. From the inoculation of would-be everyday technologies against ridicule (Michael 1997), to their integration into existing household or workplace routines (e.g. Silverstone and Hirsch 1992), such technologies – in their at once material and cultural aspects – are grounds for the making of bodies. However, at base, bodies are doubly 'made' – simultaneously constrained and enabled, 'closed down' and 'opened up'.[5] Moreover, the 'promises' incorporated within technologies are neither nec-essarily explicit, nor fulfillable in the broader context of technoscientific circulation: the configuration of bodily open/closed-ness is not always fore-seeable. For example, the telephone, originally introduced as a tool through which to conduct business (and therefore gendered masculine), rapidly become appropriated by women as a means of talking with friends and relatives and maintaining social networks (e.g. Frissen 1995). Cowan (1985) documents how the introduction of such domestic technologies as washing machines and vacuum cleaners promised a reduction in the household labour of women. In actuality, it seems to have done nothing of the sort, necessi-tating instead continued high levels of cleaning activity in order to maintain what had become higher standards of cleanliness (see also discussion in Chapter 6). Gendered bodies that should have been enabled to do other things (including rest), within the broader technoscientific circulations in which emerging technologies were tied to changing standards of acceptable cleanliness, were still required to work hard.

Nowadays, as Silva (2004; see also Lie and Sorensen 1996) argues, the role of technology (in her case study, media technologies) in the moral economy of the family is considerably more complicated than in Cowan's examples. Gendered division of domestic labour is more opaque and emergent; the complex promise of technology is likewise less obviously gendered. A contemporary example that, though arguably less gendered, echoes the patterns Cowan identifies, is that of the Dyson Dual Cyclone and now Root Cyclone vacuum cleaners. The selling point here is that the design ensures no loss of suction so that whatever is being vacuumed is cleaned at a consistent level. However, as one advert for Dyson claims, this necessitates that the bin in which dirt is collected needs to be emptied more frequently. Instead of the Dyson cleaner being used to achieve the usual levels of cleanliness (simply by moving the cleaner head more quickly), what is being advocated is greater cleanliness attached to greater activity (emptying the dirt container more frequently).

However, such mundane technologies carry with them many other implications for the body in everyday life. Let us consider the example of the electric toothbrush (Carter, Green, Wardell and Thorogood 2005). Here is a mundane technology that in some cases is still novel enough to require some domestication. However, even when well integrated into everyday routines, it requires ongoing adjustments. As Carter *et al.* document in a study of the use of the electric toothbrush in everyday life, certain bodily changes need to be implemented for these artefacts to be 'successful'. For instance, and in contrast to manual toothbrushes, the pressure with which the brush head is applied to the teeth and gums needs to be more carefully modulated in order to avoid breakage. It also requires, where the same electric toothbrush is shared (though obviously not the heads), a coordination of bodies as they move in and out of the bathroom and consecutively make use of the toothbrush. Further, and again in contrast to manual toothbrushes, several participants stated that, because of their perceived messiness, they were uncomfortable with using the electric toothbrushes in public even within their own homes. Here, we see how a mundane technology requires new comportments that ranges from the most minuscule movements, through coordination with others' bodies, to personal sequestration.

Notice how these accounts are in many ways concerned with risk.[6] On the one hand, obviously, there are risks that are putatively solved by the use of electric toothbrushes. On the other, a series of other risks are generated: risks around breakage, collective peace, personal identity and social propriety.

Everyday bodily risks and reflections

Above, we saw how a TV programme like *So You Think You're Safe* highlights the risks and dangers posed by everyday living, and how technology plays a part in these. In relation to the electric toothbrushes, some respondents were additionally sensitive to the dangers posed by the fact that these were *electric* toothbrushes. Of course, there are many other dangers posed by mundane technologies (we are routinely regaled with the 'fact' that most accidents happen in the home), but it would be well beyond the scope of this chapter to try to provide some sort of overview of these. Suffice it to say that these range from immediate dangers to the body (the ones upon which I concentrate here) to distal global risks (as Shove (2003) notes, such mundane everyday technologies are environmentally unsustainable – their impact will be on future bodies and most likely on bodies elsewhere). The theme I would like to pursue here concerns how, as the 'target' of everyday technological risk, the body is broadly conceptualized. This will serve as a lead-in to a discussion of the body's emergence in relation to more 'exotic' technoscientific products and processes.

Typically, the risks normally posed by mundane technologies to bodies are ones that open it up, that play on its leakiness, or render it more leaky, in order to enter and damage it in some way – that is, that endanger its integrity. Famously, the penny farthing bicycle posed many dangers for its riders (Bijker 1995) and it was partly in response to these that what became known as the typical bicycle came to be developed and generally accepted. Contemporary examples of risky technologies are legion: cookers, heaters, refrigerators, escalators, prams, motorbikes, mobile phones, computers, and the arrangements among these entail many ways of endangering bodies. Burning, tripping, cutting, crushing, aching are ready options in everyday encounters with mundane technology. Usually, we deal with these by using our bodies 'appropriately': through avoidance, through modifying comportments, through acquiring skills, through buying the least risky products where we can. There are also sociotechnical assemblages in place that advise us about proper bodily relations to these artefacts (e.g. in the UK, there is the Royal Society for the Prevention of Accidents, see http://www.rospa.org.uk/) , as well as aiming to standardize (the risks that attach to the composition of) such artefacts (e.g. see the work of the British Standards Institute, www. bsi-global.com, work that extends to an award-winning educational website at http://www.bsieducation.org/Education/default.php).[7]

If, in these general examples, the body is endangered by being opened up inappropriately, there are also ways in which the technologies close down the body inappropriately. Michael (2000a) documents how the car and the TV remote control, in being formulated as risky, are not uncommonly seen to 'close down' the body in undesirable ways. For example, in relation to the car, the body is imagined to be liable to becoming out of control and indulging in 'road rage' because, for instance, within the car it feels cocooned and protected. As Royal Automobile Club psychologist Conrad King put it, there is 'an artificial sense of insulation, protection and empowerment provided by the car' (quoted in *The Sunday Times*, 25 June 1995). Here, the body is encased or closed off. A parallel sort of closing off is found in various accounts of the 'couch potato', where the term is typically slang for someone who constantly watches television, with the aid of a remote control, from their sofa. Michael identifies a number of discourses in the media where the couch potato's body is seen to be at risk. Obviously it is often represented as an unhealthy and unproductive body, but it is also frequently regarded as an uncultured body (in so far as it prefers the dubious pleasures of television watching and channel surfing, closing itself off from 'proper' culture) and it is an uncivic body (or politically sequestrated body in so far as it absents itself from the public sphere).

As should be all too obvious, this brief account of the risks posed by mundane technologies has been structured by the, albeit simplistic, contrast between appropriately and inappropriately open and closed bodies. Open bodies can be represented as 'bad' (in that they are endangered) or 'good' (in that they are engaged with appropriate or proper outside cultural and material artefacts). Conversely, closed bodies can be represented as 'bad' (in that they improperly fail to engage with appropriate or proper outside cultural and material artefacts) or 'good' (in that they properly distance themselves from those artefacts that would endanger them). Here, a lot hinges on the meaning of 'in/appropriate' and 'im/proper'. For example, the couch potato might well be seen as a model of corporeal indolence and irresponsibility; contrariwise, it can also be regarded as a mode of (possibly, slacker) resistance and a tactical withdrawal from the world of meaningless work or aspirational consumption. The general point is that, with this brief reflection on the supposed corporeal risks posed by mundane technologies, we have begun to explore how the assumptions and promises attached to these technologies serve in the closing up and opening out of bodies, and thus in the complex and contested imaginings and valuations of bodies as they emerge in their relations to everyday technoscience. In the next section, we begin to consider how bodies are performed, and modelled, in biomedical technoscience.

Biomedicine's bodily risks

Biomedicine, as noted in the introduction to this chapter, has been a source of many of the ways in which we perceive, understand and treat our bodies. Needless to say, these biomedical versions are considered historically specific. As Synnott (1993) documents, in the West, over the course of time, the body has been represented in a number of ways. For some traditions in Ancient Greece, the body was glorified (e.g. the Cyrenians, and to a lesser extent the Epicureans), for others the body was regarded as the tomb of the soul, where the soul was held to be superior to body others (e.g. Orphism, Plato). For the

Romans, again the body was seen as a little corpse, clay, corruption. This is partly reflected in Christianity: Paul and others, such as Augustine, regarded the body as both a temple and enemy, at once bruised and honoured. This led to a split in the early church between ascetics (with their privileging of martyrdom, virginity, celibacy, mortification, hardship, etc.) and 'moderates' (for whom the body was a revelation of God's goodness). In the Renaissance, body and soul were less at odds, but also the body began to emerge as private and secular (partly brought about through increasing mobility leading to greater egocentrism). This privatization of the body was subsequently accelerated by Calvinism and Puritanism. From Descartes on there is an emphasis on the body as machine (a dualism in which body and soul were held to be radically distinct; of course, there have been anti-dualists of various kinds, including Spinoza and the romantics). This dualistic version of the body finds expression in a variety of academic disciplines, psychology, biomedicine and, of course, the social sciences (though it has always been critiqued from the margins – e.g. Shapin 1991). That this model finds its way into popular culture is relatively uncontroversial. As a particularly extreme example, we can point to the surgeon Christian Barnard's popularizing book *The Body Machine*, which takes the mechanical metaphor of the body to heart by drawing parallels with the car (it has chapters entitled The Chassis, On the Road, Body Maintenance, In the Workshop).

Such models were tied to the various practices of medicine as they have emerged over the eighteenth and nineteenth centuries. From the perspective of Foucault (2003), the body in medicine has been, since early modernity, the subject of detailed classification and regulation. In the case of the classification, this has been rendered possible through innumerable technologies that have increasingly enabled the 'medical gaze' to penetrate deeper into the body (from the scalpels used in anatomy to the X-rays, CAT scans, NMR, and so on). Indeed, such modern techniques as the 'flythrough' in a virtual (computerized) body in the Visual Human Project, begin potentially to alter our perception of the body: as Waldby (2000) points out, where once it was a messy terrain to be mapped or traversed (and this goes for the 'normal' endoscope), now it is an open space to explore – a miniaturized version of space exploration.[8] Ironically, what is lost in this technique is the materiality of the body: the extreme mobility that is virtually permitted also compromises the Visual Human Project's usefulness for surgeons. A point that follows is that what is useful in the laboratory does not necessarily translate into something practicable in the clinic or the surgery. More immediately relevant is that the classification of the body as a particular sort of object inflects with technological developments (see Brown and Webster 2004). Indeed, in more recent times, the range of metaphors that have been used in biomedicine have proliferated. Thus, for example, the female reproductive system has, according to Martin (1987), been modelled in terms of a hierarchical, bureaucratically organized system under the control of the cerebral cortex and a manufacturing plant (the uterus) designed for the production of babies. Martin (1994) has also documented the changing conceptions of the immune system as they moved from a 'cold war' model (antibody goodies versus pathogen baddies) to a 'post-Fordist' model ('just-in-time' and flexible immune response to specific infective agents). Other metaphors, not least cybernetic or informational ones, have also become increasingly current, not least in relation to the new genetics (Haraway 1991; Birke 2000; Brown and Webster 2004).

If the body is multiply represented within biomedicine, we should also be wary of ascribing biomedicine an intellectual or cultural autonomy – that is, we need to be careful that we do not render it a 'citadel', as Martin (1998) has phrased it. Rather, as Martin's own research amply demonstrates, there is a complex 'circulation' of models of the body between expert and lay domains, a circulation that has partly been enabled by the shifting relations of power between these domains.

This returns us to Foucault's concern with the way that central to biomedicine is the regulation (and normalization) of bodies. According to the Foucauldian analytic, biomedicine's complex institutional links with the state enable it to survey, pathologize and 'police' bodies. Health systems, and their many specialist agencies, have measured health in, and promoted health to, the population. In the process, particular models of proper behaviour have been established and disseminated. For instance, programmes such as the cervical screening programme in the UK have drawn upon, and reproduced, discourses of responsible women (as well as accounts of the nature of the cervical smear and its accuracy) that women themselves rehearse in interview (Singleton 1993; Singleton and Michael 1993). Indeed, more broadly, women's bodies, especially as related to reproduction (e.g. IVF and now cloning), have been the subject of particularly close surveillance. Recently, with a view to mapping potential genetic-related disease, national programmes have been instituted to survey people's DNA (Brown and Webster 2004). However, as many authors have pointed out, these programmes meet with resistance – they are problematized by both users and medical practitioners.

There are two interrelated points to raise here. The first is that biomedicine is clearly not a uniform enterprise. There are variations not only in technique, but also in conception of the body, and relations to the patient. Sometimes these lead to competition between, or hierarchization among, specialisms, at other times a more or less peaceful and pragmatic coexistence is possible (see especially Mol 2002). Arksey (1998), in her analysis of the repetitive strain injury controversy in the UK in the mid-1990s, traced the way that different specialisms accounted for the existence or otherwise of RSI. Thus, orthopaedic surgeons, who hold to a typical medical model of the body as object (unsurprising, given their particular professional skills, which can entail surgical intervention especially in the repair of bones), were unsympathetic because RSI presented no detectable organic symptoms. In contrast, rheumatologists, who, in specializing in the diagnosis and treatment of joint damage, take a more holistic approach (e.g. examine work practices), were more supportive of the idea of RSI. Physiotherapists were the most sympathetic: being more treatment-orientated, they championed RSI. In part this pattern can be read as reflecting the different knowledge and practice orientation of these three specialisms. However, within the context of the health service, it can also be read as a competition for authority and autonomy in which support was garnered from differing constituencies (the orthopaedic surgeons from traditional biomedical authority; the physiotherapists from vocal groups of sufferers – many of whom were journalists).

Note that many sufferers (through their representative body) attempted to ally themselves with those specialists who reinforced, and promoted, their views about RSI. In a sense, then, they were in a position to 'choose' among their specialists. This stands at odds with the rather all-encompassing account of biomedical classification and regulation briefly presented above. In part this is a matter of analytic level: the Foucauldian focus upon discourse tends to neglect the complexity that inevitably arises with implementation (see Bowker and Star 1999). In contrast, at the ethnographic level of medicine, practice is shot through with variability, complexity and 'recalcitrance' (e.g. Pickering 1995). Even such latter-day claims for a shift towards a geneticization of our representations of, and increasingly our practices towards, the body (where the body is understood primarily in terms of its genetic composition) are found to be more complex at closer quarters (see Rose 2005). For instance, Kerr and Cunningham-Burley (2000) show older discourses around 'blood' cohabit with accounts in terms of genetics. However, there is a social issue at stake here too. As Brown and Webster (2004) note, there is nowadays an apparent flexibility that people have in 'choosing' their healthcare, in pursuing a more or less bespoke 'lifestyle' in which are realized different (mixtures of) versions of the body – from the typically medically 'objectified' (the body treated as an object) to the alternatively 'subjectified' (the person treated through spiritual intervention or work upon their agency). As an instantiation of the rise of consumer culture (e.g. Keat, Whiteley and Abercrombie 1994), such 'healthy lifestyle' choices find perhaps one of their most extreme expressions in the increasing availability of a variety of home (do-it-yourself) testing kits, including, over and above pregnancy tests, tests for cholesterol levels, blood pressure, bacterial vaginosis, bowel cancer (blood in the stool), osteoporosis, Alzheimer's (loss of sense of smell), diabetes (glucose in the urine) and prostate cancer (measuring Prostate Specific Antigen). While these tests raise concerns about accuracy, and genetic tests in particular raise worries about the ways in which people will deal with possible future ailments for which there are no current treatments, or whose aetiology is more than just genetic (e.g. http://www.hgc.gov.uk/testingconsultation/ genewatch.htm, accessed 14 January 2004), they offer users an apparent autonomy in determining their 'lifestyles' (as well as being part of their lifestyles).

The counterargument here is that such 'choice' is actually simply another form of government over bodies, albeit one that is more dispersed. Thus, rather than being surveilled and regulated by health agencies, people now 'surveille' and regulate themselves (see Brown and Webster 2004). This is an example of a broader movement towards 'governmentality', where persons use a variety of techniques, often derived from unlikely sources such as self-help manuals, to enact themselves as citizens (Rose 1999). In the present case, what these many 'private' diagnostic technologies do is allow people to pursue particular, divergent practices of good health and proper bodies even as they gravitate around more or less standard models of health and body. On this score, then, one governs oneself through the exercise of 'freedom' (Rose 1999).[9]

However, we should not see this process of governmentality as mediated by some pure individual (and individual body). The particular individual is the result of a governmentality that maps onto a nexus of heterogeneous embroilments. As Irwin and Michael (2003) suggest, one way of conceptualizing such a nexus is in terms of the complex and dynamic 'contestations' over the meanings and practices of bodies by 'assemblages' composed of disparate combinations of actors: laypeople, experts, politicians, media, technologies. On this account, everyday bodies are enacted through, and against, a variety of entities: families and communities, allied experts and knowledges, 'friendly' media and political spokespersons, and more or less amenable technoscientific and biomedical enactments that might range from the private diagnostic technologies noted above, through to claims to authority by expert 'elites'. This can also be read as the body emerging in relation to the hybrid practices of consumption and citizenship, and is dealt with directly in the next chapter.

Throughout the foregoing account, the body has been represented as something that is 'enacted' or 'performed'[10] (as well as produced and made). This terminology needs a little unpacking. Classically, the body has been conceptualized in social science along a dichotomy (see B. Turner 1996): on the one hand there is Leib, or 'the lived body', and, on the other, Korper or 'the possessed body' – body as subject versus body as object. We saw in the case of Latour's door closer how the lived body is altered through its interactions with a particular mundane technology. We saw how the 'possessed body' has been variously represented in biomedicine and drew attention to some of the struggles over those representations. However, as Mol and Law (2004; also 1994) point out, this leaves a bifurcation in the modes of knowing the body – either from within or without. For authors such as Burkitt (1999) the 'solution' is to transcend this by making the body 'open' so that we become aware (and can represent) its 'possessed-ness' – that is, its object-ive doings and happenings. From the perspective of Mol and Law, such a re-thinking would reproduce the emphasis on knowledge, when it should shift focus to the *doing* of the body. As they put it, 'We all have and are a body. But . . . we also do [our] bodies. In practice we enact them' (2004, p. 45). However, as they trace the many enactments (which include measuring, feeling, countering and avoiding) by which people do hypoglycaemia entail many entities (e.g. food, blood glucose measuring kits, friends and relatives): as such, the body should be considered porous or semipermeable.[11] There is a throughput of materials by which hypoglycaemia is managed, though this is not necessarily a coherent process. Indeed, this issue of coherence is further complicated by the fact that hypoglycaemic bodies also do many things other than hypoglycaemia. Coherence is an accomplishment or, rather, doing bodies is a matter of non-fragmentation. As Mol and Law argue:

> So long as it does not disintegrate, the body we-do hangs together. It is full of tensions, however. These are tensions between the interests of its various organs; tensions between taking control and being erratic; tensions, too, between the exigencies of dealing with diabetes and other demands and desires. In the day-to-day practice of doing bodies such tensions cannot be avoided. Like it or not they simply must be handled. (2004, p. 57)

What this example highlights is that the everyday body must be continuously enacted as a more or less coherent entity. Crucial to this process are the many little technologies that populate everyday life (e.g. home heating and air-conditioning systems; cleaning and car technologies) that enable this enactment, and that 'presuppose' its openness and distributedness. Moreover, as hinted above, the parameters of fragmentation/coherence are constantly subject to contestation.

In this section, we have considered the enactment of the everyday body, not least as mediated through biomedicine. We have seen how it contributes to the making of particular bodies, even as it presides over the proliferation or multiplication of bodies. In relation to everyday life, this proliferation is re-doubled. In everyday life, we are, obviously enough, assailed by many images of the 'body-now' (the variety of ways in which the body may be currently enacted), but we also routinely encounter various versions of body-as-it-will-(or might)-be. In terms of corporeal coherence, we find that technoscientific claims abound that radically alter both the way that such coherence might be enacted and, indeed, the meaning, or necessity, of coherence.

Beyond bodies: the transhuman and the posthuman

So, enacting the everyday body is conducted against a backdrop of representations of future bodies – that is, the imagining of everyday bodies of the future. As we shall see in Chapter 7, technoscience (or, rather, its particular spokespersons) constructs the 'future' in numerous ways as it goes about performing itself in the present. Technoscientific statements are littered with promises and warnings, projections and scenarios that serve in the process of swaying public opinion, mollifying regulatory authorities or attracting venture capital. In all this, the future body features centrally: images abound in which the body will be supplemented in innumerable ways, or where it will be transformed fundamentally, or where it will even be transcended. At this point we find that the lines between fact and fiction become blurred; more precisely, claims and counterclaims about what is factual and what fictional (fantastical even) are plentifully evidenced (e.g. Michael and Carter 2001) as debates proceed over the morality or efficacy of these future developments. However, clearly the future is not singular, it can be more or less far away in the western view of time (see Michael 2000c) – future bodies might be 'just around the corner' or 'way off in the future'. Moreover, a particular future body's status might alter: for example, the xenotransplanted body (a human body with replacement organs derived from animals, indeed a body for which there is a limitless supply of replacement genetically modified animal organs) was at one time (in the late 1990s) seemingly imminent, whereas at the time of writing (2005) it seems, for many, to be far off in the future, if not an impossibility (a fiction) (see Michael and Brown 2002).

Though I have only scratched the surface, the foregoing hints at the range of future bodies on offer, as it were. In this section, I will focus primarily on the more extreme or radical versions of the future body available to us in 'serious' scenarios. In particular, I will draw on two terms that have been applied to such future bodies: the transhuman and the posthuman. The transhuman can be defined in various ways, but key elements include the idea that humanity, through science, technology and rationality, will develop beyond its present form to a higher physical and mental level. Of the various changes that are envisaged, near future developments will enable life extension and bodily improvements through genetic modification and bionic implants; in the longer term, the use of nanotechnology will halt ageing altogether, consciousness will be fully uploadable, and human and artificial intelligence will be integrated (see, for example, http://www.aleph.se/Trans/index.html, accessed 19 January 2005). Obviously, such projections entail the transformation of what it means to be human – indeed, we go beyond the human, becoming posthuman. These are themes that crop up regularly in the media (as well as more specialist journals and discussion groups), but also have been a long-standing concern in fictional genres, ranging from such classics as Frankenstein to modern TV programmes such as the *Star Trek* family of TV series, and films such as *Total Recall, Terminator,* and such body horror films as *Videodrome* and *eXistenZ* (see Graham 2002; also in relation to performance art, see Mackenzie 2002).

For some writers such prospects are deeply disturbing. In Fukuyama's (2002) view, the classic liberal possessive individual is held to be grounded in the 'nature' of humans, a nature that would be corrupted by transhumanizing technologies. Without such a nature and humanity in place, the grounds for ethics are removed and moral and social coherence endangered. This contrasts with views such as those of Hayles (1999) and Muri (2003) for whom the posthuman that emerges in the context of information and communications technologies fundamentally rests on assumptions (and also traditional narratives) about the possibility of transcending the body, of seamlessly moving information across different media (corporeal and machinic). In contrast to this material seamlessness, Hayles wants to recover a more complex figure of the human – a dispersed and fluid version in which the limits brought about by embodiment are celebrated rather than seen to be matters of transcendence and transgression.

In both cases it is held that new technological developments will offer humans something radically new: if not outright transcendence, then a profound expansion of the body. In comparison, following Latour (1993), Waldby (2000) argues that posthumanism should be thought of as a series of sociotechnical developments that throw into relief the existing posthumanity of humans – that is, their everyday embroilment with more or less mundane technoscientific artefacts. Indeed, it is through such multiple corporeal connections that we enrich life, according to Latour (2004b): on this view, the opposite of embodiment is thus not such a version of posthuman 'transcendence', but death.

Be that as it may, the discourses and representations that celebrate potential embroilments with new technologies do raise issues around 'what it means to be human'. As we saw in the preceding chapter, figures such as Haraway's cyborg address the ambiguities that arise with the sort of prospectively 'corporeally *closer*'[12] mixing of the human and nonhuman that future genetic, biotechnological, bionic, ICT, nanotechnological applications imply. While they promise more or less dramatic 'improvements', they also come with risks attached. Such risks are at once material (e.g. mundane bacterial infections at the point where flesh meets plastic or metal or, more controversially, retroviral infection), cultural (e.g. the transgression of the 'purified' categories of human and nonhuman, judgements over the intent behind posthuman innovations that merely strive for more of the same – more intelligence or increased corporeal longevity) and, of course, political (struggles over benefits and marginalizations that come in the wake of these posthuman technologies and their warrants).

This section has touched briefly upon some of the issues raised by projected developments in technoscience as they apply to, and prospectively transform, the body. These representations of possible future bodies are subject to debate, debate in which, variously, there is a reassertion of the 'liberal' body, or advocacy of a limited enhancement, as well as celebration of transcendence. I have suggested, though not illustrated empirically, that these representations (and the debates surrounding them) are available in everyday life. Certainly people encounter such representations in the media, and are able to address the issues raised by these in interview (Wardell 2001) and presumably in mundane conversation. However, in the context of the possibility that everyday life can prise open the future (Lefebvre's moments of presence), these texts serve as ambiguous resources in the critique of the present and the formulation of a 'better future'. To be sure many 'future bodies' are spectacular (in the situationist sense). Yet, in so far as they embody corporeal futures that might or might not be 'better', they at least have the virtue of enabling a discussion of the meaning of 'better'. When this is allied to the 'critiques' that arise in relation to the everyday body (as discussed above), we can see that the nexus of technoscience, everyday life and the body entertains a daunting complexity. In the next section, we approach this complexity through the consideration of a putatively common corporeal feature of everyday life or, more pertinently, as Dorothy Smith (e.g. 1988) puts it, everyday-every night life, namely sleep.

Enacting sleep

In his sociological reflections on sleep and health, Williams (2002) very usefully catalogues the sorts of sociological issues that can be attached to the 'topic' of sleep. Reworking Williams' analysis through the structure of the present chapter, we can begin to unravel some of the technoscientific dimensions of sleep. As Williams notes, sleep is not an unproblematically 'natural' bodily state: it is also a resolutely cultural event. As Williams (2002, p. 178) puts it, 'Sleep ... is a complex, multifaceted, multidimensional phenomenon, which is irreducible to any domain or discourse' but incorporates the biological, psychological, environmental, structural and the sociocultural. As such, we should not be surprised to find that there are divergent body techniques for 'doing' sleep (including standing up, on horseback, in short bursts of 'power napping'). Moreover, such enactments of sleep are enabled by a variety of consumer goods and services that promise to enhance the body – making it more healthy or more beautiful (Williams and Boden 2004).

Following on from this, like other mundane corporeal events sleep is performed in relation to certain 'invisible' technologies. Typically, in the

West, patterns of sleep have developed in the context of such innovations as the use of bedrooms (the gradual sequestration of sleeping), changing forms of lighting (and acceptable levels of illumination), as well as other modes of comfort (e.g. temperature control). These are, of course, not determining factors: as Shove (2003) illustrates, the waning of the siesta can be related to the increased availability of air conditioning, but also to the need for co-ordination with the 'normal' (9 am to 5 pm) working hours of other countries.

Like Latour's door closer, the technoscientific paraphernalia of sleep shape the comportment of bodies. Most obviously, the mattress is an artefact that influences the way in which bodies 'get comfortable'. For example, in addition to the different internal design of mattresses (e.g. pocketed spring, open spring or latex), there are also foam/polymer coverings that soften in response to body heat and thus mould themselves to the body's shape. Some mattresses therefore claim to adapt to bodies, indeed, some double mattresses can be sprung to two different tension levels so that each sleeping partner has the mattress softness that best suits them. Of course, these are mattresses at the upper end of the market, and many who cannot afford these have to adapt their bodies to their existing mattress (e.g. change postures) or supplement the mattress with other technologies (e.g. extra bedding between mattress and body – such as the Nimes mattress).

At stake here, paralleling the discussion above, are issues of risk. Mattresses are sold on the basis of the buyer's need to get a 'good night's sleep': without such a mattress one is at risk of not getting such sleep, and thus at risk of all the ailments and accidents that putatively follow from sleep deprivation. As Williams argues, sleep has become an object of medicalization (he notes in his review of the literature that western societies are routinely declared to be 'sleep sick'). Remedies include medication, relaxation techniques and alternative therapies. Thus, we can detect a transition towards what has been called healthization, in which there is ongoing surveillance of the body by the self as it pursues particular 'lifestyles'. Getting a good night's sleep becomes a matter of planning, and organizations such as the Sleep Council in the UK provide 'sleep tips' such as 'keep regular hours', 'create a restful sleeping environment', 'make sure your bed is comfortable', 'take more exercise' and 'cut down on stimulants' (http://www.sleepcouncil.com/consumer_room/sleep_tips_2.cfm, accessed 16 December 2004). To follow these practices (of governmentality) is, in part, to recruit various technoscientific artefacts that act in relation to the openness of the body as it performs its coherence: technoscientifically enabled self-surveillance resources the everynight doing of sleep and thus the accomplishment of non-fragmentation (instead of the fragmentation that would putatively follow with sleep deprivation). Needless to say there are many such artefacts that can flow into this process. For example, in relation to sleep disruption through snoring, we find such over-the-counter technologies as Snore Calm Chin Up Straps, Air Plus Nasal Strips, Snore Calm Foam Ear Plugs, Snore Calm Herbal Spray, True Sleep (a device that applies an electric pulse to the arm and induces the sleeper to turn over, thus stopping their snoring). Moreover, there are other strategies for dealing with sleep disruption too (e.g. moving rooms) that, as Hislop and Arber (e.g. 2003) trace, are deeply gendered – that is, serve in the process of gendering. In all this, the simple point is that the self-surveillance of the sleeping body is partly mediated through technoscientific artefacts of various sorts that, by virtue of the body's openness, allow for the management of its cohesions (or non-fragmentation).

However, this all assumes the necessity of 'a good night's sleep'. Micro-sleeps – episodes of sleeping that last from a few seconds to a few minutes – punctuate our days without our noticing. While these are often seen to be risky (say, in driving), they are also part and parcel of maintaining non-fragmentation, not least in relation to everyday sleep deprivation. As noted above, Williams documents a range of sleeping body techniques that continue to evolve in light of the sociotechnical possibilities provided by, for example, 24-hour TV or web-based gaming or online chatting. Instead of diurnal cycles of sleep, it is not difficult to envisage the emergence of sleep cycles that map in complex ways onto international date lines as people coordinate their chatroom conversations. In some ways, then, the necessity of 'a good night's sleep' is rather more problematic than it would at first seem; indeed, in the context of such contemporary technoscientific possibilities, it can be said to be an artefact of what, to borrow from Williams, might be called the 'sleep industry'. Now this does not deny the 'need' for sleep in general (nor, indeed, the environmental usefulness of the mass getting of 'a good night's sleep'), but it does render problematic the idea of the body's 'need' for a 'a good night's sleep'.

The foregoing embeds the idea of the necessity for sleep in the complex circulations of technoscience (and indeed, to echo Haraway, the new world order – see Chapter 2) that serve in the performance of the body as at once opened up and closed down. However, there are also accounts which argue that with the aid of technoscience it will be possible to dispense with sleep altogether or partially in the everyday future body (see Williams 2002). Transhumanists, for example, both applaud recent developments such as the antisleep pill (see http://eric.transhuman.org/transhumanism/, accessed 21 January 2005) and advocate the transcendence of sleep through the pro-spective uploading of consciousness (e.g. http://www.betterhumans.com/ Better_Health_through_Democratic_Transhumanism.Article.2003-06-02-3.aspx, accessed 21 January 2005). Here, then, the body is placed within a techno-scientific lineage that progressively rids it of its deficiencies and limitations: accordingly, there can be no luxuriating in slumber (Williams and Boden 2004), only mediated wakefulness. Ironically, such scenarios reproduce the self-governing individual: shot of the encumbrances of the sleeping body, the self can really realize itself.

Concluding remark

In the previous section, I used an example of everyday (every night) bodily activity, namely sleep, to illustrate the overall analysis of the complex role played by technoscience in the constitution of the everyday body. We have traversed a number of sleeping bodies: from the 'invisible' constitution of the comfortable sleeping body, through the consuming sleeping body, through the medicalized sleeping body to, in the end, the unsleeping non-body. In the process, we have cycled through a consideration of automatic practices, more reflexive actions, co-constituted actions with experts of various ilks (both medical and alternative), to prospective becomings. The body has been conceptualized as something to be 'enacted', but this doing, which is, generally speaking,[13] fundamentally concerned with non-fragmentation, is attained through its openness, not least its openness to technoscientific artefacts. What holds the body together, then, is, partially at least, a throughput of technoscientific materials and representations that bind – that is, enable the performance of – its 'parts' into a more or less coherent whole, even as those parts relate in different ways to different technoscientific materials and representations. Of course, this throughput is a manifestation of relations of power – some bodies are more privileged than others in the variety and quality of such a technoscientific flow. Further, the 'wholeness' – or non-fragmentariness – to be attained hardly reflects some abstract norm; rather, it is the ongoing product of deeply social and political processes that are folded into the prospect, emergence and availability of various technoscientific artefacts (e.g. new vacuum cleaners, new mattress materials, new medical technologies and procedures).

As noted previously, such dynamics of coherence and fragmentation, multiplicity and unitariness are tied to patterns of doing politics – a doing that nowadays straddles the everyday enactments of 'citizenship' and 'consumption'. It is to these enactments and their inflections with technoscience that we turn in the next chapter.

Notes

1. Here, the term 'biomedicine' is being used to denote a subset of technoscientific circulations that are concerned with what might be called 'healthy bodies', though, as we shall see, these bodies are highly variable, deeply contested and chronically multiple.

2. Indeed, if the body is conceptualized dynamically, as ongoingly responsive to its environment, then the border between the normal and the pathological is highly variable and, by implication, the relevance of biomedicine is always in prospect as the norms of normality shift (see Canguilhem 1994).

3. The Disability Discrimination Act 1995 subsequently had added a revised Code of Practice that addresses the duties placed on service providers in relation to the physical features of their premises. From 1 October 2004, it is required that service providers must make 'reasonable adjustments' to their premises to overcome physical barriers to access.

4. Having noted this, what such strategies do is reinforce the notion of the autonomous individual who, of course, is mediated by the collective processes of legislation and implementation. Ironically, one possible downside of such strategies is the social 'disabling' of other actors in knowing how to respond to disabled people: the technology does the job that bystanders once did, or could have done.

5. In Whitehead's (1929) terms, the body emerges as an actual – atomized, or rather singularized – entity; but this is on the basis of the concrescence – or the 'coming together' and 'combination' of various prehensions (which include both the material and semiotic). As such, the body is both 'brute fact' and 'processual'.

6. It is very easy, as the present volume illustrates, to deploy the terminology of risk given the prevalence of such notions as risk society or risk culture or, more generally, of risk discourse. However, 'risk' has its own peculiar origins and we might wish to be wary of the way that risk connotes calculability, or balancing. Instead, we might wish to speak of 'problems in living' or, even, 'avenues in living'. The gentle point is that, for all its contemporary utility, we should not fetishize the term 'risk'.

7. Indeed, the BSI advertises its 'Kitemark' by noting that 'Consumers and specifiers trust that products bearing the Kitemark have passed a rigorous certification process and as such will not only be safer to use but will also be fit for the purpose for which they were designed' and that, as such, the 'Kitemark offers peace of mind to our clients by reducing the possibility of product recall as well as increased sales by product differentiation' (http://www.bsi-global.com/Kitemark/Product/index.xalter, accessed 10 November 2005).

8. It should be obvious that it is not being suggested that such innovations lead to greater accuracy in the representation of the body. As Fleck (1979) noticed long ago, anatomies selectively highlight some features while impoverishing others. This is further underlined by the fact that some bodies serve as the templates for others: the male body has traditionally been the prime model for the human body, with the female body regarded as simply a 'lesser' version.

9. Mol (2002) beautifully details a number of ways in which different enactments of arthrosclerosis within a hospital are resolved or left pragmatically

diverse. However, under the market-ish conditions outlined here, irresolution is of the essence as each practice equates to something like a market niche.

10. While I use the terms performance and enactment interchangeably, Mol (2002) wants to privilege the latter, on the basis that it does not carry the complex connotations of 'performance' – a term that has a long and convoluted history in the social sciences, humanities and arts.

11. Note that the 'openness' of the body in this account is in terms of an ontology, rather than, as in the discussion of the couch potato, in terms of morality.

12. 'Closer' in the sense that such technologies work not only on the surface of the body, but within, through and around its flesh.

13. I say 'generally speaking' because people also do their bodies as dying, as fragmenting. This is most obviously the case in instances of euthanasia. For an account of the debate around euthanasia, not least around the issue of 'dignity' – which can here be read as a particular form of performance – see Hausmann (2002).

4 Technoscientific citizenship: the micropolitics of everyday life

Introduction

In this chapter we trace a path through the relations between technoscience, everyday life and politics. Within the critical tradition, everyday life is a domain structured by various mechanisms through which people's capacities are 'constrained', or their potential dissipated. Surveillance and consumption are key motifs in this respect: we are surveilled so that we better govern ourselves; we are everywhere encouraged to consume in order that we do not ask difficult questions of the state, or capital, or the new world order. Here, we underline the key role of technoscience: it is fundamental to the many modes of surveillance, and it is instrumental in the turnover, distribution and signification of consumables. However, it is also possible to rescue the motifs of surveillance and consumption in order to show that the everyday is not as devoid of 'politics' as the critical perspective seems to imply. Further, technoscience seems to play another role in the politics of everyday life: technoscientific institutions of one sort or another have often aspired to monopolize both the questions and the answers to various issues (e.g. matters of biomedical or environmental risk), not least by attempting to dictate the terms of reference, the forms of argumentation and the standards of evidence (see Barry 2001) for addressing such issues. In other words, it is not an uncommon impression that everyday debates, criticisms, interventions of various sorts are systematically corralled by technoscience. And yet, as we shall see, technoscience and its institutions are hardly monolithic – they are perfused with the public in its various guises, and indeed increasingly 'invite' laypeople to contribute to the process of technoscientific decision or policy making.

To illustrate, let us consider a mundane example: supermarket food shopping. This is a highly surveilled activity, and it is obviously concerned with consumption. Sidestepping the many complexities of shopping (e.g. Miller 2001), here I wish merely to point out how shopping for food in a supermarket might be considered as something like a political act. Shopping, like so many everyday activities, is embroiled in the multiple processes of surveillance – processes in which technoscience is central by serving in the production of 'panoptical' spatialities in which everyday life is disciplined. Moreover, technoscience serves in the collection of data about the shopper so that he/she can be targeted for future consumption. At another level, the role of technoscience in the production of consumer goods, and in particular in the seemingly accelerating innovation (or new-ness) of such products, can be said to be instrumental in the sort of consumer 'passivity' that Lefebvre bemoaned (see Chapter 2).

However, technoscience also contributes in contrary ways. In the simple act of reading, say, an ingredients label – something that many of us do on a regular basis – and deciding not to buy a particular brand because of high salt levels in the food, lack of labelling regarding genetically modified foodstuff content, fair trade or the amount of E-numbers, we do 'technoscientific politics'. Our consumption choices are here resourced by a form of surveillance: we surveille producers, and our consumption choices can have political effects (for instance, somewhere down the line, on technoscientific institutions charged with regulatory duties). So, in this brief example, we find 'countersurveillance', 'political consumption' and 'civic involvement' all rolled into the seemingly simple process of supermarket food shopping.

This illustration, of course, disguises a wealth of complexity (not least in how such 'political' consumption operates in relation to companies rather than the nation state, where the latter is perceived as unwilling or helpless to act). As we shall see, the roles of technoscience here are deeply variegated, and it is difficult to tease them into any recognizable pattern. Nevertheless, the politics of technoscience crucially entail the making and self-making of what we might call 'technoscientific citizens'.

The aim of this chapter is thus to begin to unravel the complex role of technoscience in such everyday 'political' acts. If we reconsider the preceding example, we can get a further sense of this complexity. Technoscience constitutes a topic of politics: the GM content, the E-numbers are technoscientific artefacts that need to be surveilled. Technoscience is also a resource in that it feeds the particular concerns about foods: it is (a faction of) technoscience that judges this or that artefact to be more or less risky. And technoscience also partly comprises a medium – the label is a technical invention (that is constantly being reinvented to deal with changing transportation, storage and retail conditions), as are the various media through which warnings are circulated, and the rise or decline in purchases of a particular product registered by retailers and, ultimately, producers.

In sum, technoscience is at once a topic, a medium and a resource for the doings of everyday political actors. The processes and products of technoscience are subjected to political scrutiny; technoscience contributes to the media by which old political alliances (or what we shall later call 'assemblages') are sustained and new ones become enabled; technoscience provides resources (knowledge, spokespersons, practices) for political intervention.

Let me briefly elaborate on the above in terms of citizenship. As we saw in Chapter 3, technological artefacts are central to many everyday enactments (e.g. tooth cleaning, sleeping) with their characteristic micropolitics. The micropolitics associated with such artefacts incorporate a 'wider' field of politics in which those artefacts have been developed, produced and assessed in various ways. Crucially, publics regard the relevant technoscientific institutions with deep ambivalence – a coalescence of profound scepticism and chronic trust. This is linked to an increasing porosity between technoscience and public (or science and society) in which politics does not entail some sort of stand-off between expert and lay constituencies, but a mixing up of these constituencies in hybrid assemblages.

One dimension of this wider politics of ambivalence entails the mutual scrutiny of 'science' and 'society'. If putative representatives of public actors such as non-government organizations keep a close watch on technoscientific institutions, those institutions have shown a keen interest in those publics. For instance, they have been keen to stress the role of scientific knowledge in making 'good' citizens; despite many critiques of this perspective, it remains alive and well (Gregory and Miller 1998). However, there is also, as part of a broader democratizing dynamic, a movement towards 'engaging' politically with the public. Technoscientific institutions now go about facilitating the involvement of publics in participatory or deliberative fora in which publics supposedly make a contribution to the process of scientific decision or policy making. This process at once assumes and facilitates particular versions of the public, and, indeed, of everyday life. Everyday life becomes the domain of technoscientific reflection and 'responsibility', and the inhabitants of that everyday life a constituency ('stake holders') to which technoscience must explain itself – render itself transparent. As we shall see (and in a further exemplification of the porosity of science and society), the techniques of making transparent draw upon the techniques of doing self in everyday life.

In what follows, then, we begin with a consideration of the ambiguous role of technoscience in the 'politicizing' microroutines of everyday life. We then expand on this to suggest that, to the extent that there is a pervasive sense of risk, technoscience can serve at once as focus, medium and resource for that sense and its political expression. In particular, I point to the complex emergence of hybrid assemblages along whose fissures politics might be said to be enacted. Entailed in such assemblages is the production of particular types of technoscientific citizens. That is to say, in so far as technoscience is embroiled in, and an instantiation of, the 'democratization of democracy', it reformulates everyday life as a particular sort of political domain populated by 'technoscientific citizens' that must be enabled in one way or another. In conclusion, these various political moments are summarized and retheorized around the notion of ethno-epistemic assemblages.

Technoscientific micropolitics

This section explores the ambiguous role of technoscience as it at once contributes to the conservative micropolitics of everyday life, while simultaneously opening up possibilities for political intervention.

As we have already seen, technoscience in the form of mundane technologies, both old and new, serves in the routinization of everyday life, a routinization that recreates existing relations of power, not least those around gender. Thus, vacuum cleaners, microwave ovens, remote controls, and so on, despite the more or less explicit promise of a reconfiguration (that is, 'liberation') of everyday life, reproduce its gendering. However, as we have also seen, this is no one-way process. These technologies also serve as opportunities: they can be reappropriated in unexpected ways that undermine or reconfigure domestic relations of power. For example, Lally (2002), in her broadly ethnographic account of the introduction, and assimilation of, computers into the home, documents how, depending on the circumstances, this complex technology affirmed gendered divisions of labour or served in undermining them. On this score, Lally's study contributes to the view of gender as something that, in the context of domesticating technologies, is emergent. As we noted in Chapter 2, and as Lie and Sorensen (1996) insist, what is needed is a 'shift away from viewing gender and technology as predetermined' (p. 20) – in the process of domestication, both gender and technology are negotiated.

Another example of the way in which the role of a technology impacts upon everyday relations of power can be found in the doctor's surgery. A visit to the general practitioner has changed in many ways over the last 20 years or so. The paternalistic relation between doctor and patient has increasingly been supplemented, if not outrightly superseded, by a more service-orientated model in which the doctor is the service provider and the patient is the consumer (e.g. Fairclough 1992). This shifting relation is further mediated by various technologies, not least information and communication technologies, which afford patients ready access to a variety of medical information (though this is hardly an unproblematic blessing – laypeople are well aware that they need to treat such information with circumspection – e.g. Michael and Carter 2001). However, such technology, and specifically the apparent movement towards computerizing patient records (as electronic patient records (EPRs)), might also work towards reasserting old relations of power. The hard-won right of patients to access their records might be eroded by the rise of EPRs as patient-held records become but a pale imitation of the EPR. As Brown and Webster (2004) put it, 'In practical terms, the advent of the EPR shifts these boundaries of ownership over records away from patients and back to the clinic again' (p. 93).

This latter example brings into relief one of the key ways in which technoscience impacts on the micropolitics of everyday life, namely surveillance (in the above case, of the body and patient behaviours). De Certeau's account of the strategies of discipline in everyday life hinged on a view of a spatiality characterized by chronic surveillance. This spatiality has, arguably, been hugely expanded through technoscience. In a sense, then, everyday life, is gridded by lines of sight enabled by a multiplicity of technologies (or, better still, sociotechnical assemblages) – the EPR is one such technology, but more obvious still are surveillance cameras (of various sorts) and computer-mediated surveillance (not least of working patterns and personal consumption patterns (see Lyon 1994, 2001)). Inevitably, there are other techniques on the horizon. Such prospects include: fingerprint and retinal scanning; microchip tracking devices (generically called radio frequency identification tags – RFIDs – tiny enough to be integrated into products such as clothing); and genetic surveillance through which genotype is determined with all the ramifications that follow in the wake of such information (for personal relations, labour, insurance, healthcare).[1]

In the rest of this section, I want to explore some of the ways in which everyday micropolitical complexities proliferate around technologically mediated surveillance. I begin with the brief example of buying a new computer, before moving on to traffic speed monitoring systems. When it comes to buying a computer, an obvious option is to visit one of a number of large retail outlets – in the UK these might include Currys, Dixons or PC World. The choice of store already entails a certain exercise of political discretion. As Lally (2002) notes, some of her respondents reflected not only on the sort of computer they required, but also on the sort of retailer with whom they wanted to do business. For some, it was more important to go to small local retailers partly because of their perceived commitment to after-sales service, partly because of a desire to support local business and thus retain the particular character of the locality. Even at this initial point of inquiry, there is a form of surveillance of retailers. On the internet this surveillance is even more pronounced as different retailers are given averaged satisfaction ratings derived from users' comments and judgements (e.g. http://www.ciao.co.uk/). As Castells (2001) traces, everyday surveillance is caught up in a series of ironies, especially that of countersurveillance in which the surveillers are themselves surveilled.[2] Here, then, entering into a highly surveilled environment such as a megastore can entail a tacit 'calculation' between something like 'low cost' and 'locality and service'. Such calculations have, inevitably, been resourced by the more or less close scrutiny and 'surveillance' of the ICT retailers.

The surveillance systems within the space of a megastore are not panoptical: there are always spaces that remain unseen. Partly this is because there are simply not enough cameras to cover the whole store. This renders the idea of surveillance through the panopticon (through which everything is seen by an unseen seer) problematic. An alternative model might be that of the oligopticon (Latour 1998; Boyne 2000) through which a little is seen well. However, even this 'little seen well' is subject to ruses: unseeable spaces are created through bodily movements, positionings and shapings. Moreover, as noted by Ball (2003), those who do the surveilling are themselves 'situated' – their surveillance patterns differ depending on their relative interests, background, prejudices, and so on. This latter observation is important in that it points to the way that social and technological actors mutually shape each other: the surveillance equipment is directed, scanned and interpreted in ways that are always limited by human capacities just as those capacities are expanded by that equipment. Put another way, everyday life is not surveilled by abstract entities but by concrete humans operating within what is their own everyday life (which itself is likely to be under surveillance – see below). Here, then, De Certeau's (and Foucault's) schema of strategic space needs to be rethought because the practices of surveillance are mediated through the vagaries (and, indeed, 'ruses') that run through the everyday lives of the surveillers, as well as the surveilled. What seems to emerge in this example are two disjunctive everyday practices each with their own orderings and disorderings: that of everyday shopping and of daily surveillance. Onto these practices are projected by theorists such as De Certeau two divergent idealizations: the everyday concrete space of tactics and the abstract space of surveillance.

Above, it was briefly suggested that those doing the surveillance – security guards, in this instance, but as we shall see this extends to many actors, such as the police and scientific institutions – are themselves open to surveillance. On the one hand, this is an internal matter as targets are set, activities are audited and practices rendered transparent; on the other, such organizations are themselves open to scrutiny by publics (as we have seen in terms of their quality of retail service) but also through various more or less sustained surveillances of the 'surveillance society', both academic and institutional. In the UK, the most obvious institutional actor in this respect is the Information Commissioner's Office – an independent body whose mission is the promotion of 'public access to official information and (the protection of) your personal information'. Inevitably this organization is itself subject to surveillance as set out in its 'Transparency Policy: Disclosing Information about Specific Indi-viduals and Organisations' (see http://www.information commissioner.gov.uk/eventual.aspx).

What all this suggests, in contrast to the models of a panopticon or oligopticon, is a dispersed and mutual form of surveillance: surveillance and countersurveillance operate recursively, mirroring one another in an unfolding (or enfolding) pattern. Perhaps a more useful motif here is that of the taleidoscope. The taleidoscope is a version of the kaleidoscope. It shares with the kaleidoscope a tube along which at least two or more rectangular mirrors are situated. However, instead of the kaleidoscope's translucent bead-filled compartment, the taleidoscope has a lens that permits views of the world beyond. With this arrangement, various lines of reflection symmetry generate complex, seemingly fragmentary, patterns of outside scenes, which, when the tube is rotated, are shifted and changed. Within the taleidoscope, images of the 'world' (its surveillance) – a world partly populated by others surveilling the process of surveillance through their own taleidoscopes – are being reflected and re-reflected in complex and shifting patterns in which any original point of view (that is, image) becomes impossible to identify. To the extent that this holds any water, we might tentatively put forward, as a counter to the panopticon and the oligopticon, the heuristic model of the 'taleidoscopticon'.

Let us illustrate this idea of the taleidoscopticon further, this time in relation to another prominent mode of surveillance, especially in the UK: that of speed cameras on roads. Any road user will be aware of the roadside furniture that cohabits with cars and drivers: the signs and barriers whose everyday aim is the ordering of driving. Entailed in such bits of roadside furniture are assumptions about the nature of the driver. For example, Stenoien (1994) traced the evolution of the model of the driver in Norway from 'uneducated' but essentially rational (in the pre-1970s) to a more criminal figure post-1970s, which possessed bad attitudes. Where the former needed more information (in the form of more traffic signs), the latter was an object of normalization and surveillance (e.g. through speed cameras).

Judging by the proliferation of speed cameras, and the range of speed camera innovation, in the UK the 'criminalized' driver model certainly holds sway. The logic of their widening introduction is ostensibly driven by a desire to reduce road accidents (HM Office of Science and Technology, Postnote, May 2004, No. 218), and their siting is dependent upon a number of criteria related to accidents (e.g. four collisions resulting in death or serious injury over three years and within one kilometre for the installation of fixed cameras). However, recent changes in policy that have allowed local authorities to use funds derived from speeding fines to finance more speed cameras have led to a widespread suspicion that the real reason behind speed cameras is the making of money for the Exchequer (£20 million profit in the year up to April 2004, (*Observer*, 13 February 2005)).[3] This suspicion of a 'stealth tax' is further underpinned by the view that the cameras are often sited in locations that have less to do with securing safer driving than maximizing fine income for speeding violations. Not unexpectedly, these views are disputed by the UK Government (HM Office of Science and Technology, Postnote, May 2004, No. 218).

Of course, such surveillance meets with countersurveillance measures that take a variety of forms. For example, there are attacks upon the cameras – the destruction of speed cameras by shooting, chainsawing down or burning with rubber tyres – which 'reorganize', albeit temporarily, the space of surveillance. Until the recent changes in funding policy, based on reports in the press, we could imagine that there was awareness of the financial straits faced by police forces, which meant that they were unable to afford to develop speed camera film. Here is another example of the everydayness of this oligopticon's operators, which is in this case surveilled through the media.[4] There are also websites devoted to speed cameras (notably, in the UK, http://www.ukspeedcameras.co.uk), which document changes in government policy, differences in police force implementation, the location of speed cameras for various districts, and features on various aspects of speeding (for example, a section devoted to the excuses given by allegedly speeding drivers: having a poorly budgie and rushing it to the vet; desperate for the toilet and speeding to the nearest public convenience; picking up a nameless and addressless hitchhiker and letting them drive the car because they liked it). More pertinent are the reviews of, and advertisements for, various techniques for avoiding detection or escaping prosecution. Among the former are a number of technologies designed for use against various sorts of traffic cameras, such as GPS speed camera warning systems for Gatso cameras. In addition, there are web pages devoted to means of refuting the speeding charges made by the police (including links to a website – http://www.pepipoo.com/ – which in aiming to empower drivers provides information on the law in theory and practice, and examples of real speeding cases) and links to a further website advertising the book *UK Driving Secrets* (http://www.uk-driving-secrets.com/beat/menu.aspx?afl=10531), which, written by an ex-policeman, claims to be able to help drivers escape fines by pointing out various loopholes.

These 'countersurveillance' measures serve in reordering the everyday surveillance spaces of the 'traffic speeding detection system'. They are part of a broader configuration – what I've tentatively called the taleidoscopticon. Of course, in line with this configuration, we would expect such countersurveillance processes to be counter-countersurveilled. For example, the UK Government is in the process of possibly banning devices designed to detect, and forewarn drivers about, upcoming speeding cameras. Moreover, there are other non-governmental actors engaged in promoting more surveillance. For example, BRAKE, a road safety charity in the UK, wants more surveillance to bypass the countersurveillance measures mentioned above. Thus, in campaigning, BRAKE wants 'road safety enforcement to be brought into the 21st century'. For BRAKE, the

> advanced technology is available, in equipment such as Automatic Number Plate Recognition (ANPR) cameras, mobile roller brake testers and portable computers with licence-holder information, but the resources are not there to allow this equipment to be rolled out nationally, and in numbers to enable effective enforcement throughout the UK on a daily basis.

It is this lack of surveillance that BRAKE wants to remedy (see http://www.brake.org.uk/index.php?p=273, accessed 24 November 2005).

In this example we get a hint of the complex loops of surveillance that the notion of the taleidoscopticon is designed to capture. This is indeed merely a hint, as it is easy enough to follow the spirals of surveillance in many directions, most obviously in relation to the use of the internet. Further, the everyday technoscientific politics of the taleidoscopticon can be seen to be bound up with contestations over rights (the various and conflicting rights of drivers, not to say pedestrians) and risks that are both physical (e.g. endangerments of fast driving) and social (around the very issues of surveillance). As we have seen, these contestations entail actors that are heterogeneous – minimally comprised of laypeople (drivers), experts (lawyers, police) and technologies (camera detection devices, the internet).[5] In this section, we have engaged with a variety of micropolitical technoscientific practices that can be called citizenly. In the next section we go on to explore the technoscientific micropolitics of everyday life, but this time in specific relation to the figure of the technoscientific citizen.

Technoscience and the three heterogeneities

In this section we shift focus a little to explore further the relations between technoscience and everyday politics. The main point to make at the outset is that these relations are characterized by (at least) three heterogeneities that in various ways reorder the associations between, for want of better terms (though these have a long lineage), science and society. These heterogeneities entail, variously: the de-differentiation of science and society (or more broadly the expert and the lay); the erosion of the distinction between citizen and consumer; and the blurring between the (rhetorical) registers of 'emotion' and 'rationality'. The aim here is to unravel further some of the complexities that characterize what we have been calling technoscientific citizenship.

The de-differentiation of science and society

In Chapter 2, we encountered, in Lefebvre's version of everyday life, a modern society marked by increasing differentiation. Thus, according to Lefebvre, there was a progressive separation of various domains of social activity, such as, for example, the divergence of economic and political 'systems' from the everyday world of lived experience. Particularly relevant here is the differentiation between, broadly speaking, science and society (e.g. Nowotny, Scott and Gibbons 2001; Irwin and Michael 2003). The proliferation of expert institutions and scientific disciplines as the social and natural was increasingly analysed and rendered calculable and controllable distanced science from the layperson, though this was not a smooth process by any means (Crook, Pakulski and Walters 1992). Tied up with this development of science was the promise of progress, political as well as intellectual and technological. However, as numerous theorists have argued, science has reneged on this promise; it is now seen to be a servant of big business – it is technoscience, in Haraway's terms (see also Lyotard 1984). According to this narrative, then, there is a chronic disparagement of science.

In contrast, this apparent erosion of science's status can also be thought of as a symptom of the de-differentiation of science from other spheres, not least everyday life[6] (Nowotny et al. 2001). After all, the early modern differentiation (or professionalization) of science from other sectors of society was a huge historical accomplishment (e.g. Whitley 1984; Shapin 1991; Gieryn 1999) by particular scientists and groups of scientists in the process of institutionalization; there is nothing natural or inevitable about it. In the face of a number of 'driving forces', as Nowotny and her colleagues put it, the boundaries between science and society are blurring – there is a 'co-evolution'. Of these 'driving forces', two seem particularly relevant.

The first is the 'growth of uncertainty'. Accordingly, 'both science and society have opted for the production of the "New" in an open-ended process of moving toward a plurality of unknown futures' (Nowotny et al. 2001, p. 35). In other words, the production of new knowledges, technologies and associated systems generates a sense of the future that is increasingly hard to pin down (also see Chapter 7). Further, the knowledge that is particularly relevant to everyday life (often called regulatory science) is increasingly indeterminate (it is what Funtowicz and Ravetz (1993) dub 'post-normal science'). For example, the advice on whether to use the MMR vaccine (the combined vaccine against measles, mumps and rubella), whether certain forms of nano-technology should be developed, how new reproductive techniques should be regulated – all of these are hugely complex issues that, despite attempts at simplification, cannot be reduced to technoscientific decision making. Simply determining which expert knowledge to prioritize entails admixtures of political, social and cultural, as well as technoscientific, judgement.

Second, new forms of economic rationality are emerging in which speculation is central: here 'insubstantial promises' are made and disseminated, promises that are difficult to evaluate but that nevertheless serve, not least via the media, to interest venture capital, or capture the public imagination, or curry regulatory favour. These two 'forces' (which are not accorded any causal potency) suggest an everyday world in which people routinely encounter technoscientific uncertainty and/or adventure.

This can be recast in the language of risk, trust and ambivalence. One of the corollaries of the modern differentiation of science was that it aspired to, and in many ways accomplished, authority. It was the authority that could pronounce on the nature of nature. However, in the world of uncertainty portrayed above, the products of science – its promises and its open futures – are increasingly difficult to assess. The promise of, say, a potential innovation such as xenotransplantation (the infinite supply of animal organs and tissues for transplantation) becomes a risk when the possibility of retroviral infection is raised. Moreover, the very process of promising becomes risky: resources that go into funding potential innovations such as xenotransplantation open up certain futures, but close down others (as we shall explore further in Chapter 7).

Notice, however, that it is often technoscientific expertise that serves in the identification of such risks. As Beck (1992) points out, without scientific institutions going about their routine expert business we would almost certainly remain unaware of very many problems that have come to public attention. Indeed, such environmental conditions as ozone depletion, climate change and increasing ionizing radiation levels are undetectable without technical investigations and scientific studies. As both Giddens (1991) and Beck (1992) note, laypeople are caught in a relationship that is variable, shifting and unstable with regard to scientific institutions. Irwin and Michael (2003) sketch out the following crosscutting and sometimes contradictory patterns of trust. There is a chronic, systemic and unreflected upon dependence on such institutions (routine standardization and safety testing). There is a virulent distrust of, and disillusion with, them (when things go wrong). And there is a pervasive but provisional trust in them (as problems are identified and solutions offered and implemented). In other words, laypeople in the contemporary West display an 'ingrained ambivalence' towards expert systems.

Needless to say, this willingness to question the authority of technoscience by aligning oneself with technoscience takes various forms. For instance, Epstein (1996, 2000) traces how AIDS activists adopted and adapted scientific knowledge (not least around the design of clinical trials), in the process themselves becoming expert and, ironically, endangering their very status as laypeople legitimately belonging to the AIDS movement in the USA. As we saw in the previous chapter, in Arksey's (1998) account of the RSI case in the UK, laypeople can recruit, or form alliances with, certain specialisms or professions within biomedicine. In these examples, technoscience is a resource in the critique of technoscience as a topic. Technoscience enters into the everyday, and the everyday enters into the technoscientific. Here, then, citizenship is played out in a complex and shifting process of differentiation and de-differentiation from technoscience in the form of experts of one sort or another. There is consecutively and simultaneously impermeability and porosity between science and society.

'Beyond' citizen and consumer

It is a commonplace among social theorists to note that contemporary social life is characterized by an increase in the prominence of consumption. People's sense of who they are is partly conditioned by what and how they consume (rather than, say, what they produce or how they work). For example, Lash and Urry (1994) have argued that key to contemporary society is what they call specialized consumption, where individuals emerge in 'disembedded lifestyle enclaves' that are the targets of niche marketing. Instead of identification with class or community, people enact themselves in terms of the core 'lifestyle' features of their relevant enclaves, not least the consumption patterns that characterize those enclaves. Of course, what is being consumed here are not just material objects that perform some function or other (clothes, cars, computers, etc.) but the signs that are inseparable from such objects (and services) – signs that express 'the sort of person one is'. In other words, as Featherstone (1991) has put it, everyday life has become 'aestheticized' (or stylized), through which everyday life is typified by 'the rapid flow of signs and images . . . consumer society . . . confronts people with dream-images which speak to desires, and aestheticize and de-realize reality' (pp. 67–68). The upshot of this is that consumption is as much concerned with attaining images as material things. Goods are now used as if, as Lury (1996) phrases it, 'they were works of art, images or signs, to be engaged with via processes of fantasy, play, daydreaming and image-making' (pp. 77–78).

In the present context, consumption is of interest because of the ways in which it inflects with the process of doing politics – that is, of being a 'citizen' or engaging in what might be called 'citizenly activities'. Inevitably, this is a highly complex and varied process, as we shall see. The main point, however, is that in the context of technoscience and everyday life, as many authors have noted, consumption and citizenship can no longer be so unproblematically distinguished, or so readily located respectively to the separate spheres of economics and politics. As Keat, Whiteley and Abercrombie (1994) note, since the early 1990s the rise of the 'New Right' has led to a focus upon the rights, choices and decisions of 'the customer'. Domains that had once been regarded as outside the field of consumption (e.g. public services) began to be, and still are, treated in terms of the market. Organizations, such as the health service in the UK, became increasingly subject to market forces and orientated to consumer action (as opposed to citizenly intervention). This section reflects on some of the ways in which consumption and citizenship are heterogeneously interwoven. Indeed, along with the previous and the next section, we can begin to form a picture of an emergent hybrid political actor that populates western technoscience and everyday life.

Now, as Gabriel and Lang (1995) document, there are a great many different ways in which the divide between consumer and citizen can be blurred, from consumer/citizens 'voting' with their purchasing choices to making concerted efforts to influence policy making with regard to the production of certain consumables (such as genetically modified foodstuffs). Michael (1998a) has addressed the rise of consumer culture specifically in relation to the field of 'public understanding of science',[7] arguing that commitments to studying and enfranchising the scientific citizen have to be supplemented by an awareness of the role of consumption in politics. For example, science – notably in its popularized forms of books and documentaries – is a consumable good that confers particular status. Not so long ago, within certain more or less elite subcultures, not to have read (or at least started) Stephen Hawking's *A Brief History of Time* would have mildly subverted one's social standing or cultural capital (Bourdieu 1984). Such status is part of the armoury of everyday micropolitics. If this form of technoscientific consumption equips persons 'cognitively' for everyday micropolitics, there are also ways of preparing people 'corporeally'. As such, technoscience can be used as a means to aestheticize 'bodily consumption' wherein one or one's offspring can be biotechnologically redesigned better to meet the rigours of everyday micropolitics. For instance, Jill Turner (1996) notes that with the arrival and ready availability of recombinant human growth hormone, it was used not only to increase the growth rates of children who presented with unusually low levels of endogenous growth hormone, but was also used on 'short normals' (shorter children with the expected levels of endogenous growth hormone) in order to improve their 'life chances'. In both cases, we might say that technoscience has served to resource actors (intellectually or corporeally – but both culturally) in the micropolitics of everyday life.

This version of micropolitics might seem to be at some remove from the notion of citizenship – a notion that typically connotes some form of political relationship with the state. However, here we operate with a much more dispersed version of the 'political'. As the governmental literature has amply shown, government of citizens is increasingly devolved to individuals themselves, who must 'work upon' themselves in order to govern themselves (e.g. Rose 1999). The resources for this process of 'working upon oneself' are multiple and derived from disparate sources (e.g. psychological and social scientific disciplines but also, increasingly, biomedical and technoscientific – see Rose 2005). What we encounter in such instances, then, is the government being self mediated through acts of consumption that are, in one way or another, enabled or resourced by technoscience, that serve in the making of citizens – a form of government that takes place in everyday life.

So, in relation to technoscience and, specifically, aspects of biomedicine, this dynamic can be thought about through the interplay of discourses of rights, and wants and needs. If consumption can be seen in terms of the fulfilment of 'needs' or 'wants', as Slater (1997) unravels, these needs and wants can be matters of enormous contention, not least because there are innumerable criteria by which to assess whether something is a need or a want. However, the obvious point I wish to draw out here is that there is a fine line between what is a want, a need and a right – a fine line whose length and width is partly shaped by technoscience (in its guise as a sociotechnical assemblage in which circulate myriad texts and materials). Returning to the example of recombinant human growth hormone – what was a matter of need (supplementing the shortfall of endogenous hormone), becomes a matter of want (that is, becomes routinely available to short normals as enabled by the biotechnologically mediated abundance of the hormone), which in the particular social context (competitive social life) comes to be rearticulated as a need and a matter of right. The point here is that a technoscientific innovation (recombinant biotechnology) has generated a 'solution' that, in a particular social and cultural setting, seeks out further problems that, in becoming routinely solvable (enabled by the assemblage), undergo, albeit contested, a transformation from wants to needs to rights. This sort of pattern (which is by no means technologically determined, but technoscientifically enabled and sociotechnically contested) can be detected in all manner of innovations – body modification innovations (it is a right to have breast implants), *in vitro* fertilization (it is a right to have children) and, most recently, pre-implantation genetic diagnosis and sex selection (it is a right to have a balanced ratio of boys to girls in a family – see Dickens 2002; *Guardian Unlimited*, 27 October 2005, http://www.guardian.co.uk/international/story/0,3604,1601160,00.html). In sum, the availability of given technoscientific innovations within a particular assemblage can resource particular claims to rights – the rightful access to new procedures or treatments. Arguably, it is the struggles over the 'right', or otherwise, to some technoscientific innovation that comprises much of contemporary western politics and bioethics (at the cost of more sustained attention to other global 'rights' to medical treatment for such widespread diseases as malaria – see Rose 2005).

Of course, the interplay of needs, wants and rights within technoscientific consumption, and their relation to the micropolitics of everyday life, can also find expression in more usual citizenly interventions (e.g. lobbying, pressure group and social movement politics). In all this, rather than disentangling what is 'citizenly' and 'consumerly', it is more fruitful to trace out the ways in which these conflate – or rather, hybridize – in citizenly-consumerly practices. Though this heterogeneity is hardly surprising, by tracing controversies over the role of technoscience in everyday life, we can begin to unravel the citizenly-consumerly practices through which what is seen as a matter of bodily necessity in one environment can be regarded as profligate luxury in another.

One of the features of these contestations is that media exposure to disasters 'elsewhere' in the world – disasters such as the Boxing Day 2004 tsunami or the Pakistan earthquake or the US floods – can be deployed to puncture rhetorically particular claims to what is 'technoscientifically rightfully ours'. The role of technoscience under such circumstances comes to be aligned with, for want of a better term, human rights (to shelter, food, etc.). The failure of western technoscience to deliver aid can in itself be a matter of citizenly-consumerly practices, not least around charity giving. Other technoscientific failures can also serve to relativize certain 'technoscientific rights' (to sex balance, for instance) – failures of dams, of nuclear facilities, of containment of disease can be used to shift debates, assert more pressing 'needs' and insist upon the 'rights' of others. Having noted this, the images of technoscience and technoscientific failure can also become spectacles to be consumed. On this score, they can be used to re-enact national identity (say, around mourning). This is simply the other side of the consumption of technoscientific gee-whizzery in which advances are celebrated, most especially if it can be established that 'home-grown' scientists had a hand in this discovery or that breakthrough. Governments are not above reminding their citizens of the excellence of national scientific and technological research and design (witness the nationalistic fanfares around Nobel Prize winners). Conversely, the consumption of such technoscientific spectacles becomes aligned with the enactment of political identities or imagined community – whether local, regional or national.[8]

Above, I hinted at the way that technoscience, as a terrain of public controversy, might also be a spectacle to be consumed. In the conduct of such controversy, claims to rationality are – or have been – critical. Technoscience is, after all, a key bastion of the western traditions of rationality, and debate and deliberation over technoscientific matters might be expected to take rational forms (e.g. in relation to animal experimentation (see Birke, Arluke and Michael, forthcoming)). And yet, such deliberation, and its representation, are perhaps changing to incorporate more 'emotional' dimensions. Indeed, as I will suggest below, the politics of technoscience is drawing on everyday affective practices. This reflects the third heterogeneity in which 'science' and 'society' blur, in which 'emotion' and 'rationality' mix in possibly novel ways. We turn to this complex issue in the next section.

Mixing 'emotion' and 'rationality'

In the previous sections we encountered some of the ways in which the relationship between scientific expertise and laypersons has become more pliable. Laypeople as citizens engage scientific expertise in a number of ways that can blur the distinction between 'science' and 'society'. Moreover how, and the ends to which, they engage technoscience cannot simply be characterized as 'political'. 'Technoscientific citizenly' activity is, I have suggested, now interwoven with acts of consumption – acts that also serve in the production of knowledge. Indeed, as Irwin and Michael (2003) have argued, it is perhaps better to envisage the relation between publics and technoscience in terms of heterogeneous mixtures of actors – laypeople, scientists, but also journalists, activists, social scientists and regulators, to name a few. They call such combinations 'ethno-epistemic assemblages': a heuristic conceptual tool through which to investigate the shifting interrelations – the differentiations and interminglings – between science and society. In brief, the term 'ethno' connotes the idea of locality such that knowledge is always produced and taken up in the context of local cultural conditions – in a word, situated. The sociological tradition of ethnomethodology (Garfinkel 1967) is also evoked by this term in order to emphasize the fact that social activity can be understood only by reference to where and when it occurs, and thus knowledges that emerge from those activities are always, in principle, contestable. 'Epistemic' is meant to invoke the production of truth or, more accurately, truth claims, thereby drawing attention to the fact that the sorts of assemblages in which we are interested are fundamentally orientated towards the production and distribution of claims about what is real, whether this concerns science, politics, ethics, economics or human experience and identity. The final term, 'assemblage',[9] taken from Deleuze and Guattari (1988), refers to the collection of various heterogeneous fragments that can entail 'territorialization', which we here characterize as stabilized (or rather, reiterative) patterns of relations (that generate particular differentiations between, say, science and public). Such territorializations can be disturbed and the differentiations that characterize them subject to breakdown: the term de-territorialization evokes this process of mixing up (intermingling or blurring). Further, this intermingling can 'settle' into new patternings of relations – a re-territorialization becomes established in which new differentiations pertain.

Such a framework, when related to technoscientific citizenship, allows us to pose such questions as: What are the patterns of differentiation and intermingling between science and lay?[10] What is the range of knowledges (that might draw upon aesthetics, ethics, politics and culture as well as science) that can contribute to de-/re-/territorialization? What makes up the array of practices, materials, discourses that constitute technoscientific citizenship? Ideally, ethno-epistemic assemblages allow for the exploration of the means by which publics (and experts) construct, reinforce and blur the boundaries between science and society in various ways and, in the process, articulate (and perform) their citizenship in at once routine but also unexpected ways.

In the present instance, I am interested in tracing out one particular facet of ethno-epistemic assemblages, which points to a third heterogeneity in which what counts as a persuasive discourse within technoscientific politics has shifted. Increasingly, I want to suggest, it is not only arguments that are ostensibly grounded in scientific rationality that are rhetorically potent: discursive performances that are 'emotional', specifically that enact a revelation of suffering, are becoming an increasingly prominent part of the discursive arsenal of technoscientific citizenship and politics.

This can be linked to the previous discussion of the role of consumption. Arguably, the consumption of 'emotion signifiers' has become a major characteristic of contemporary western societies. Such 'emotion signifiers' – that is, overt emotional performances – have gained currency through a range of developments, not least what Nikolas Rose (e.g. 1996, 1999) has called the 'psy complex' (comprised of the knowledges and practices of such disciplines as psychology, psychiatry and psychotherapy), with its conventions, discourses and techniques of self-revelation and self-governance. Such techniques (or the results of such techniques) have pervaded everyday cultural forms, not least the audience-participation or reality TV shows such as *Oprah* and *Big Brother*, as well as, of course, soap operas. Specifically, we can suggest that the very act of being open and displaying one's private suffering, pained emotions, cognitive anguish and so on, and reflecting upon such suffering, pain, anguish and so on, serves to enact the 'limits' or 'core' of self. As such, these displays of suffering are markers of 'authenticity'. This performance of suffering not only appeals 'politically' or 'morally', but also 'aesthetically' – indeed, as suggested by the countless examples of TV interviewers asking their interviewees how they 'feel' about some state of affairs or a recent (dramatic or tragic) event, it is almost as if the display of, and accounting for, emotion (and especially suffering – the camera almost invariably zooms in on the face with its quivering chin and watering eyes) is demanded. Here, then, the display of spectacular emotion becomes 'credible' on, for want of a better term, aesthetic grounds (e.g. Lash 1988).

In the context of citizenship and technoscience, everyday arguments over the rights and wrongs of a particular technoscientific innovation are hardly couched simply in 'rational' terms but have a prominent 'emotional' component. Indeed, it is a standard complaint that publics are far too 'emotional' when it comes to such issues as animal experimentation (e.g. Michael and Birke 1994; Michael and Brown 2005). However, as Irwin and Michael (2003) note, more nuanced uses of emotion displays are possible. For instance, people will argue against a technology such as xenotransplantation on the basis that, despite their suffering and the possibility of being beneficiaries, they cannot countenance the biomedical use of animals. Here, suffering (not only that of physical pain, but also in relation to the decision-making process in which one must balance potential benefits against animal suffering) is performed by being articulated or discoursed, and thereby used to authenticate a particular position within a technoscientific debate. There is no contradiction between emotion and rationality in all this: rather the supposed dichotomy between these registers collapses as 'emotions' are articulated 'rationally' and, indeed, 'rationality' is rooted in 'passion'.

This emotional genre of self-presentation also seems to be entering into the discourse of spokespersons for scientific institutions. For instance, scientific institutions, in their attempts to regain the trust of laypeople, attempt to make themselves accountable in everyday life. As we shall see in the next section, this accountability takes various forms, not least the engagement of the lay public in participatory or deliberative events. Whatever the range of knowledges, technoscientific institutions must accommodate (whether this is exclusively expert, or also includes lay knowledge), the key point is that, increasingly, scientific decision makers must make their processes of decision making transparent.[11] However, such transparency is culturally problematic. The scepticism that prompted scientific institutions to 'open up' in the first place does not stop when those institutions start practising some variety of openness. In this context, emotion displays can serve as ready rhetorical resources that make such decision making 'authentic' and credible. Thus, in reporting through the media, spokespersons for scientific institutions enact the struggle they have had to go through in reaching a decision – the pain they have had to endure in accommodating disparate viewpoints and knowledges. As Brown and Michael (2002) note, sufficiency of difference (of viewpoints) does not rest on transparency per se (which can be implausible), but on *accountable pain*. That is to say, being transparent does not mean that the decision-making process has canvassed as widely as possible. Rather, what makes the array of positions taken into account in the decision-making process acceptable is the suffering of the decision maker. The underlying rhetorical form runs thus: *many, many disparate points of view have been accommodated, this has been a painful process and all that can be borne has been borne.* Given that such emotion displays are themselves rhetorically potent, a decision is rendered plausible by virtue of the ostensible 'pain' or suffering the decision-making body (instantiated in the spokesperson) has had to endure, rather than on the basis of the propriety (transparency, integrity, participation, consultation) of the decision-making process.

To the extent that the above account holds water, we might expect that emotion displays, as an everyday rhetorical currency, have begun to pervade technoscientific politics. As such, taking a position on a technoscientific matter of concern becomes mediated, in part, through emotion displays: one's position is rendered 'plausible' or 'credible' through the ostensible fact that one has suffered.[12] If such is the stuff of everyday technoscientific micropolitics, the exchange of emotion displays is also a matter of consumption: inseparably, one is persuaded by such displays and, because they are also 'sensational', one consumes them. In other words, over and above the role of consumption in the everyday doing of politics, politics has become the 'object' of everyday consumption.

The last three subsections have traced a number of ways in which 'science' and 'society' have intermingled to produce citizenly actors who, in their everyday lives, do not only ambivalently draw upon technoscientific expertise, but do so in ways that entail aspects of consumption and share 'rhetorics of suffering'. As such, what it means to be a 'technoscientific citizen' in everyday life becomes rather heterogeneous, and, in some ways, difficult to pin down. This, of course, does not mean that there are no attempts to demarcate such citizens. As we shall see in the next section, the technoscientific citizen is not only being studied and consulted, but in the process is being 'made'.

Populating everyday life with 'technoscientific citizens'

The character and role of citizenship in relation to scientific institutions has been a matter of considerable interest over the last 30 or so years. Lay publics are of course highly malleable entities. As already hinted at, the emergence of science as a set of distinct institutionalized disciplines went hand in hand with a reconstitution of the 'lay public' whose 'voice' in expert matters diminished (e.g. Shapin 1991; Chaney 1993; Warner 2002). In the contemporary setting where such institutions seem to be having their credibility challenged, the public is being re-reconstituted.

In the 1970s and 1980s, the incredulity of public constituencies in relation to a number of technoscientific developments (e.g. nuclear power, toxic waste management) was largely put down to the public's ignorance or scientific illiteracy. The solution was to inform the public – once they had the facts to hand then they would support technoscientific institutions (e.g. Royal Society of London 1985; Wynne 1991, 1992). In the process, they would become more practically competent in everyday life, more able to make informed decisions, more employable, more rounded members of western civilization: in sum they would be better citizens. To this end the public were studied in terms of the deficiencies in their knowledge of scientific facts, principles and procedures, most often through the medium of questionnaire studies (e.g. Durant, Evans and Thomas 1989).

However, as various commentators noted (e.g. Irwin 1995; Wynne 1995, 1996), there was a fundamental flaw with this account and approach. It was assumed that the 'facts' would be unproblematic: they would be readily and unquestioningly assimilated. Yet, it was the very status of these 'facts', and the legitimacy of the sources of those 'facts', that was being questioned. In other words, it was a matter of trust. Indeed, as various studies have shown (e.g. Layton, Jenkins, MacGill and Davey 1993; Wynne 1996), the assumption of public ignorance (and the institutional neglect of lay local knowledges) damaged still further the credibility and standing of scientific institutions. At the same time, while laypeople could readily query the believability of such institutions, they could accept and even assert the propriety of their own ignorance (Michael 1992, 1996).

As Michael and Brown (2000) have pointed out, underpinning the view of the public was a model of citizenship grounded in what Held (1987) has called developmental democracy, where the state and the agents of the state improve the political capacities of citizens – in this case by raising their scientific literacy. To be sure there were variations on this in practice. For instance, as Hill and Michael (1998) documented, a different model of the public can be deployed: the 1991 Eurobarometer questionnaire (see Marlier 1992) seemed to be as much about market research and product recognition as a measure of the 'public understanding of science'. Here, the political model of the public seems to combine both the citizenly and the consumerly. These versions of citizenship are routinely relayed to the public through the media with such headlines as 'With more than a third of the population not knowing that the earth goes round the sun, Britain could be in serious trouble' (*The Sunday Times*, 19 November 1989).

The counter version of the scientific citizen (e.g. Irwin and Wynne 1996; Irwin and Michael 2003) has focused on the role of trust, and on the ways in which scientific institutions have alienated the public. As Michael (2002) has pointed out, this version tends to hold to a more Rousseau-esque model in which members of the public are already fully equipped with citizenly capacities. While there are certain shortcomings with this model (e.g. it does not fully take into account the embroilment of public in global aspects of technoscience and consumption – see Irwin and Michael 2003), it has been more or less enthusiastically taken up by some scientific institutions[13] (e.g. House of Lords Select Committee, 2000). While there are still complaints about the public by technoscientific actors (e.g. they are too fickle, they are cumulatively sceptical – see Michael and Brown 2005), there has been a reorientation towards an 'engagement' with the public. Thus, over the course of the last decade and a half or so there has been a growing interest in how scientific decision making or policy making might be made more democratic – that is, in developing methods through which the public might contribute to the process of decision or policy making. The reasons that might be listed for this shift of emphasis are manifold, and might range from the impact of certain influential scholars through to the broader dynamics of what Giddens (1998) calls the 'democratization of democracy' (see Irwin and Michael 2003). Whatever is credited as a major influence on such developments, there does seem to be plenty of evidence to suggest that technoscientific policy making is becoming more deliberative, consultative, participatory (see Irwin 2001; Hagendijk and Kallerud 2003). There are numerous studies that are concerned to develop and test the mechanics and efficacy of particular forms of deliberative technique. Thus, for example, Abelson, Forest, Eyles, Smith, Martin and Gauvibin (2003) examine the relative merits, and the respective fitness for use, of such participatory mechanisms as citizen juries, citizen panels, consensus conferences, deliberative polling and focus groups. Such work assumes that these democratizing techniques can reflect the lay public's concerns, and make available the local knowledges of such publics to the process of deriving technoscientific decisions or policies (see Fischer 2000).

Inevitably, there are numerous criticisms levelled at these initiatives. On the most basic level, one can query the actuality of public participation in these democratic events – it has been argued that they might be better classified as a version of public relations (Beder 1999). One can ask whether the public are taken seriously or still seen to be deficient, merely a supplement that provides the subjective values, ethics, or morals to objective science (Wynne 2001; Irwin and Michael 2003). One can also question whether the publics involved are really representative of wider constituencies (Martin 1999). One can raise the issue of whether, let alone how, dissent is accommodated in these democratizing procedures, and thus whether the emphasis on consensus forecloses other more radical forms of citizenship (Elam and Bertilsson 2003). Finally, one can regret that such deliberations take place far too late in the process of technoscientific innovation: participation should take place upstream where publics might serve to interrogate the assumptions that underlie the beginnings of emerging technoscientific endeavours (e.g. around nanotechnology – see Macnaghten, Kearnes and Wynne, forthcoming).

One basic issue we have still to address is the extent to which such mechanisms do indeed feed into the policy making processes of authorities. In considering the political status of consensus conferences, Joss (1999) identifies a number of audiences, ranging from the public themselves to national governments. This is telling, for even if such mechanisms fail to enter into governmental decision making procedures, they comprise models of what it means to be a technoscientific citizen. That is to say, they provide publics with the discourses and practices through which to do (that is, enact or perform) a particular form of citizenship. Following Hacking (1986), such participatory mechanisms can be said to 'make' citizens. Indeed, this process of making citizens is part and parcel of much social scientific research in this area. The use of surveys, interviews and focus groups equip, or resource, laypeople with particular 'skills' for comporting themselves as technoscientific citizens in everyday life, from the holding of 'opinions' to 'informing oneself about and discussing the issues' (Osborne and Rose 1999; Michael and Brown 2004).[14] In other words, if the 'study of the public' (and its understanding of technoscience, or its relations of trust with technoscientific institutions) has been supplemented with an 'engagement with the public', then what finally emerges is a multiplicitous technoscientific citizen that is made and remade not only through the productions and pronouncements of technoscientific institutions, but also through the operations – the empirical and theoretical analyses – of the social sciences. Everyday life and its micropolitics are perfused not only with technoscience but also with social science.

Concluding remarks

What emerges from the various discussions in this chapter is a picture of technoscientific citizenship enmeshed in a series of contradictory dynamics in which citizenship is at once enacted in relation to specific everyday technologies, enabled in assemblages composed in part of particular technoscientific (and social scientific) factions, differentiated against and through the technoscientific institutions that pervade everyday life, formally voiced through participatory mechanisms to which such institutions are (more or less) committed, and incongruously resourced by prevailing modes of consumption, emotionality and spectacle, as well as the more usual forms of political rationality contained within numerous democratizing initiatives.

But perhaps this is the wrong way of going about accessing the multiplicity of the 'technoscientific citizen'. Politics, as represented in this chapter, has tended to be a resolutely human affair. To be sure various technoscientific artefacts and processes have littered my analysis, but they have not been integrated into the account as forthrightly as they could have been. They have enabled and facilitated people in their everyday political doings, rather than having been a constitutive part of 'technoscientific citizens'. This is hardly surprising given the lineage of the term 'citizen' with its evocations of 'the people' and 'the social' (see Urry 2000). The next chapter considers in more detail how 'society' is made in everyday life – and how, in the process, the circulations of technoscience are at once pivotal and excised.

Notes

1. As Brown and Webster (2004) demonstrate, this interlocking of genetic knowledge and information and communications technologies leads to a model of the body that can be said to be 'bioinformatic'. Here, the body comes to be characterized – that is, becomes identifiable – in terms of information that circulates electronically. In this form, it also becomes liable to increasing levels of electronic surveillance.
2. There is much to say about the use of ICTs and the internet in the doing of politics, and we shall touch on this again. However, the limitations of space mean that I cannot pursue in this book the role of the internet in the possibly changing form of 'citizenship', or the rise of the 'netizen', or the mediation of politically and technoscientifically active 'virtual communities', or the nature of 'cyberdemocracy'.
3. At latest drafting (December 2005), there are signs in the media that this system of 'funding through fining' is about to be disbanded.
4. I refer to the 'oligopticon' of the traffic surveillance system because, as Ball (2000) notes, Gatso cameras – which are the most widely used in the UK –

have a limited purview in space (direction, angle of vision) and in being spaced out they do not, by any means, see everything – that is, panoptically.

5. We can note that technoscience has served in the micropolitics of speed cameras as resource (camera detection devices; technical discourses that problematize the accuracy of speed cameras); as a topic (the whole socio-technical assemblage in which speeding cameras are embedded); and as medium (the internet sites through which many of the critical resources are communicated).

6. As noted in Chapter 3, and unelaborated here, the doings of science – both experimental and theoretical – can be informed by representations and practices from everyday life (e.g. Martin 1998).

7. 'The public understanding of science' is what Irwin and Michael (2003) call a multidiscipline in which quantitative survey measures of public under-standings of scientific knowledge vie with qualitative case studies of the lay local trust in scientific institutions. For a sense of the range of analyses that fall under this rubric, see the journal, *Public Understanding of Science*.

8. Such political identification can take 'eccentric' forms. On a personal note, my father, who was an ardent Stalinist, routinely – it's tempting to say, in-terminably – held up innovations in Soviet science and technology as evi-dence of the superiority of communist systems.

9. This is a very simplified rendering of 'assemblage'. For instance, I make no effort to follow Irwin and Michael (2003) and unravel its machinic and enunciatory dimensions, which are not directly relevant to present purposes. More germane is that such assemblages can be 'territorialized', 'de-territo-rialized' and 're-territorialized'. That is to say, their configuration – what links to what – can be routinized, unravelled, repatterned. And all these can take place at the same time.

10. Irwin and Michael (2003) are careful to insist that science and society divi-sions can be reasserted as well as eroded.

11. This process of opening up decision making can be associated with another trend in neoliberal societies, namely what Michael Power (1999) has called the 'Audit Society' in which there is an increasing drive to have institutions articulate their rationale, account for their procedures, formalize their prac-tices, and so on.

12. Brown and Michael (2002) suggest that this combination of the 'rational' and the 'emotional' is something that is not just limited to technoscientific poli-tics. However, the present analysis contrasts with Mestrovic's (1996) notion of 'postemotionalism' wherein 'quasi-emotions' have 'become the basis for wide-spread manipulation by self, others, and the cultural industry as a whole' (1996, p. xi). Rather, the account presented here is more concerned with how such emotion performances are made credible. Further, I differ from Barbalet (2001), who argues that the experience of self as a centre of emo-tional feeling is an index of the experience of lack of effectiveness in external

market and state processes. Instead, emotion displays are here viewed as a valuable political resource. Indeed, the popularity of such terms as 'emotional intelligence' in business circles, suggests that emotion displays have become part of such 'external market processes'.

13. For example, in the House of Lords Select Committee on Science and Technology's Third Report (2000), we find a critique of the very term 'Public Understanding of Science' because of its undertones of condescension. 'Science and Society' is now the preferred term, not least because it hints at dialogue. Having noted this, the report is still a deeply ambiguous document that retains an emphasis upon non-expert citizens' need and ability to comprehend aspects of science and technology (see Section 1.11, 3.11). To echo Gregory and Miller (1998) once more, the deficit model in many respects continues to hold sway, not least among the scientific, political and policy elite (at least in their public pronouncements).

14. This point can be related to the processes of governmentality in which social science plays a part in the 'making' of persons by supplying various techniques through which persons come to know, 'work upon' and govern themselves as particular types of citizen in relation to the scientific institutions. Of course, this is enormously complicated by the many other resources, not least those that feed the emotion displays mentioned above.

5 Technoscience and the making of society in everyday life

Introduction

Eating as an everyday activity is commonsensically and sociologically regarded as a social event (e.g. Beardsworth and Keil 1997). Even when the diner sits alone in front of the television, a TV dinner perched on their lap, this is no lesser a social event in so far as it is a way of enacting 'eating' that embodies particular social expectations, and attracts particular social judgements (e.g. it is the antithesis of a 'good' sociality, or is a marker of the decline of a particular version of society). The commonplace view that in the UK people work too readily through their lunchbreak (and often fail to eat at all during the course of their working day) serves as a critique of contemporary pressurized 'life-styles', not least when contrasted with what are regarded as more 'civilized' societies where 'good' sociality is mediated through the collective practices of eating. In any case, across the dining table (imaginary or otherwise), we can be said to 'do' society in the two main broad senses identified by Williams (1988): we interact with one another to produce our social relationships, and we enact society in its abstract aspect by realizing or manifesting its class or status or ethnic divisions. The sorts of social interaction (microsocial processes) and the sorts of society (macrosocial structures) that are played out across the table are intimately tied to the paraphernalia of the table. Classically, the type of cutlery that has been used has been associated with the way western societies have become more 'civilized' (Elias 1939/1994). The combination of furniture (dinner table, kitchen table, TV dinner tray, couch, chair, stool) and food (gourmet, fast food, organic, mass produced) can be said to be a marker of cultural capital (Bourdieu 1984). At another level, the sheer variety and amount of food on the table reflects the forms of production and transportation that are available to more privileged societies (Braudel 1981).

So, if the dining or kitchen table (in the widest sense) is an everyday site around which people gather and 'do' society, it is also evidently a site littered with the products of technoscience. In the everyday comportment of preparing and consuming a meal, social relations are played out, and society is, ostensibly, manifested. At the heart of this is technoscience. This can be thought of through the work of Michel Serres. As Serres (1995b) has long argued, what he calls quasi-objects are crucial to the enactment of social relations. To quote:

Our relationships, social bonds, would be airy as clouds were there only contracts between subjects. In fact, the object, specific to the Hominidae, stabilises our relationships, it slows down the time of our revolutions. For the unstable bands of baboons, social changes are flaring up every minute ... The object, for us, makes our history slow. (1995b, p. 87).

However, these quasi-objects are not separate from, or added to, human relations. Rather, '(t)he relations at the heart of the group constitute their object; the object moving in a multiplicity constructs these relations and constitutes the group. These two complementary activities are contemporaneous' (Serres 1991, p. 102). Objects, such as the 'products of technoscience' that litter the dining room table, are thus pivotal in the doing and making of society.

But this is a complex process and it is one to which Modern westerns have, according to Latour (1993, 1999), been rather blind. For Latour, the modern West has, until relatively recently, strictly demarcated the human from the nonhuman, and the social from the material. While we inhabit the world more or less happily with numerous nonhumans, old divisions are constantly reasserted (or purified, as Latour puts it). The news media do not address hybrids of human and nonhuman but report under headings such as science, politics, economics, and so on, which keep these separate. And the spokespersons of these disciplines reassure us of our traditional (dichotomous) categories – human and technology, society and nature, cultural and material – despite what Latour regards as the increasing self-evidence of heterogeneity evoked in the numerous innovations of, for example, biotechnology and biomedicine, in which what is social and what natural becomes increasingly difficult to disentangle.[1]

Let us take an example of one such a hybrid and its tricky relation to the sociality associated with the dining table. To be a couch potato is, at once, to be a part of a subculture in which social relationships take a particular form (e.g. slacker), and to personify the decline in society's ability to interest (a proportion of) the young in healthy lifestyles, responsible living and productive sociality (Michael 2000a). Technoscience enables the enactment of this figure of the couch potato in numerous ways. For example, it furnishes the mundane technologies of couch potato-ness (sofas, TV dinners and remote controls), it mediates the couch potato both as a self (couch potatoes in interaction on the web) and as the object of scrutiny (reports in news media), and it provisions the couch potato with new, more or less exotic, products of consumption (TV meals and delivery foods; versions of the smart home would be a couch potato idyll). In the context of the critique of everyday life, there are numerous ironies to be unpicked here, most obviously in the contrasts between, on the one hand, the apparent social impoverishment of the person encased in the couch potato and, on the other, the idea that the couch potato embodies its own critique of a society characterized by aspirational consumption and the 'rat race' (see Guthman (2003), for an exploration of similar ironies in relation to the consumption of organic food). Part and parcel of these contrasting valuations of the couch potato as it goes about its everyday business are representations of the good society (e.g. one in which people can fulfil their potential as workers, as culturally active members of western civilization, as responsible health pursuers) or of society at risk or in decline (e.g. dissipated by acts of leisurely dissipation).

This chapter is thus concerned with the ways in which the complex and multiplicitous circulations of technoscience in everyday life serve in the making of 'society' or, to borrow from Law and Urry (2004), in 'enacting the social'. Needless to say, this will be a disparate process that will involve, as evoked in the above example, attending not only to how 'social' (or rather, heterogeneous) relations are enabled and configured, but also to how representations of society – particular accounts of the social world – serve in that enablement and configuration. While we will, obviously enough, be paying particular attention to the roles of technoscience in all this, we will also consider how, in the 'making of society', technoscientific artefacts (and other nonhumans too) are separated out, even rendered invisible. As we shall see, sociology itself has played no small part in such 'purifications' of the social.

In what follows, we begin with a brief consideration of the way that people talk about society in everyday life and the sorts of functions such talk might have. In this way, we begin to trace the complex interplay between technoscience and 'society'. We show how the social interactions in which talk takes place cannot be separated from technoscience (qua mundane technology), even as the divide between technoscience and society is routinely enacted. Indeed, as noted, the products of technoscience serve in the mediation of social intercourse in such a huge range of ways that it is perhaps better to think of those interactions as heterogeneous or hybrid events. Such events entail both the 'simple' co-presence of objects like furniture and cutlery, and complex mediations of technoscientific systems that serve as the semiotic/material conduits (or channels) for communication (e.g. information and communication technologies). Depending on which sort of artefact is analytically pursued, different societies come into focus. Paralleling the multiplicities we encountered in relation to the body, we find a multiplicity of societies, too. We explore such issues in the context of collecting 'social data', and in the use of the Post-it® Note and ICTs to trace out society. However, in the everyday, doing of society technoscience is itself a focus of discourse – a topic – not least when it is relatively novel. In this regard it is, not uncommonly, seen to 'drive' society in particular directions, in the process generating risks or problems. Associated with such scenarios are particular characterizations of society: it can be a risk society, a biosociety, a network society, a surveillance society, a geneticized society. To be sure, such terminology (along with the technoscience) is performative and, as such, might facilitate the 'making' of particular versions of society. However, we shall especially explore the relation of the *proliferation* and *multiplicity* of such 'society' terms to the doing of society in everyday life. Of course, such multiple representations of society entail more or less sophisticated representations of technoscience itself, especially regarding the ways in which scientific knowledge is produced and expertise accredited. On this score, we need to bear in mind that technoscientific expertise, and the everyday practices of scientists, operate within their own 'society' in which 'facts' and artefacts emerge out of local processes of negotiation, processes affected by, going full circle, the ways in which technoscientists represent society and everyday life. We can bring these various strands together through a consideration of the three TV series that fall under the rubric of *Crime Scene Investigation*[2] – a leading example of a genre that we might term the 'technoscientific police procedural'.

Talking about, and doing, society

In a study of the discursive use of terms such as 'social' and 'society', Bowers and Iwi (1993) identified eight models of society that respondents used rhetorically to 'either legitimate the respondent's argument or undercut an opponent's' (p. 368). For instance, society was held to be 'uniform and total': all people were held to be members of a particular version of society as internal differentiations were downplayed. The rhetorical utility of this account was that it allowed the speaker to render a controversial issue like pornography as unproblematic by equating it with society as whole and thus all people. Alternatively, society could be represented as 'opposing the speaker' – a rhetoric that rather than undermining the speaker's position highlights the deficiencies of 'society', which is unsurprisingly felt to be immoral or oppressive. For Bowers and Iwi, such talk is performative in so far as it functions to, hopefully, shape listeners. Accordingly, these respondents are offering 'a network of associations in their arguments, a set of alliances between actors intermingled with associations between actors and states of affairs, values and beliefs ... which if found convincing and acted in accordance with could be(come) society' (p. 389).

But of course such talk always takes place in settings seething with 'nonsocial' objects. Across the dining table, as people discuss the nature of society, various mundane technologies must be organized or disciplined in ways that allow talk to proceed smoothly and uninterruptedly. But this is not simply a case of creating a space for social intercourse; mundane technologies also 'carry' such talk by enabling the enactment of a particular self or serving in the signification of a certain status. So, in contrast to Bowers and Iwi's analysis of discourses of society, we focus upon the way that such discourses are embedded in heterogeneous arrangements of humans and nonhumans, specifically mundane technologies.

As Latour (1993) and others have noted, sociology itself is no slouch in the routine obfuscation of the nonhuman in the production, and purification, of 'society'. If we take seriously Giddens' (e.g. 1984) notion of a double hermeneutic in which the concepts of sociology (and the social sciences more generally) contribute to the pool of everyday concepts that people have ready to hand (concepts such as class, ethnicity, the unconscious, community, opinions and attitudes) and these, in turn, affect the conceptual and empirical work of sociologists, then obviously a key concept that circulates within this hermeneutic is 'society'.[3] To get at society, sociologists have recourse to social data (in the broadest sense) as evidence of the specific social events of phenomena in which they are interested, and that contribute to the making of 'society' in general. Social data – say, interview transcripts, ethnographic reports or social statistics – have to be derived in the first instance through the recording of 'social events', whether these be seemingly 'ordinary' (e.g. 'naturally occurring conversation') or 'extraordinary' (e.g. action research interventions or focus group sessions). But how do such social events and the data that they 'yield' come to be seen as 'social'? As we have argued, such settings are, after all, cohabited by humans and nonhumans. The next section examines problems in the research process in order to illustrate how technoscience features in the local making of 'social data' and thus of 'society' – a society made up of both social interactions and large-scale organizations. In other words, it traces the impure purification of society in which social data is rendered social, purified of its nonhuman and technoscientific content partly through the very 'interventions' of nonhumans.

The impure purification of 'society'

Michael (2004a) considers the example of what he calls the disastrous interview episode in which he managed to derive, seemingly, no discernible social data. In 1989, while engaged in fieldwork into the public understanding of science, Michael conducted a number of interviews with local residents in the Lancaster area. He focuses on one interview, the second of two with an ex-drug user, who, after a period of unemployment, had recently got a job at Burger King. He describes the interview layout: he was seated on the sofa, the respondent was in an armchair to his right, and the tape recorder was placed on the floor between the two of them. The interview was designed to explore the respondent's views on nearby nuclear installations, including Heysham nuclear power station, the Sellafield nuclear reprocessing plant and the nuclear submarine shipyard at Barrow-in-Furness.

During the preliminary conversation, the respondent's pit bull terrier entered the room and sat on Michael's feet; the respondent reassured Michael that the dog did this because 'she liked to know where people were'. As Michael tried to get the interview going, it became apparent that the respondent preferred to discuss her new job at Burger King and, in particular, the new possibilities this offered, not least of which was, she said, the prospect of rapid advancement. As she talked about this, her cat also entered the room and, after playing with the tape recorder, started to drag it along the floor by its strap. In the process it moved outside of the recording range of what had subsequently become a monologue about the career structure, and the imminence of promotion, at Burger King. As Michael recollects, he was too distracted by the pit bull's proximity to rescue either the tape recorder or the interview (for which five pounds sterling was paid).

This is at once an everyday social event (a conversation within which are played out certain relations of power) and an extraordinary sociological event in which no social data are 'gathered'. It is ordinary because of the co-presence of 'pets' as well as various more or less mundane technoscientific artefacts (sofa, tape recorder); and seemingly extraordinary because of the apparent breakdown in communication in which the apparently common task at hand (the interview) was undermined by one of the participants. However, as will become apparent, these ordinary and extraordinary moments are thoroughly interwoven and instrumental in making – that is, purifying – the social. On this score, the social scientific interview serves as both an exemplar of society as social intercourse and a resource by which such society is technically 'made' and circulated by social scientists.

Michael presents three analyses of the disastrous interview episode. In his first analysis, he notes how the process of communication between interviewer and interviewee was disrupted by the cat and the pit bull terrier. Their various activities materially intervened in the process of social exchange between interviewer and respondent. For example, the pit bull terrier, by sitting on Michael's feet, meant that he was too distracted to pursue the interview schedule and guide the interviewee towards the questions in which he was most interested. Part of this distraction derived from the meanings that the pit bull terrier had for Michael: at the time stories about 'devil dogs' that were capable of unprovoked attack abounded in the media – the pit bull terrier was identified as particularly liable to indulge in such an attack. Michael suggests that this analysis brings to the fore how various nonhumans must be subjected to discipline – kept in their place – for social data to be possible. Keeping the 'pets' out of the way is just one such form of discipline. Before data can be 'gatherable' – all manner of potential disruptions (or 'parasites', as Michael, drawing on Michel Serres (1982b), calls them) must be kept in check;

these can include children and neighbours, radios, televisions, washing machines on spin cycles and partners drilling into an adjoining wall. Indeed, such social data is impossible without such processes of exclusion. In other words, for it to be possible to have social data, there is a need to discipline a range of heterogeneous relations.

Michael's second analysis points to the role of the 'pets' in enabling the respondent to talk about Burger King so assiduously. The activities of the pit bull and the cat are part of the sociality of the household and as such can be seen to contribute to the 'ease' with which the respondent 'goes off at a tangent'. Indeed, they can be seen to be mutually constitutive in so far as in the context of their routine interrelations, the 'pets' act as they do because of their 'owner', and, crucially, vice versa. Michael suggests that these relationalities between pets and owner suggest that together they can be viewed as a singular entity (what he calls the 'pitpercat' made up of pit bull terrier, person and cat). He suggests that it is this entity (or co(a)gent – see Michael 2000a)[4] that disrupts the relation between interviewer and tape recorder that together (as another co(a)gent – an intercorder) generate and record social data. Of course, the range of constituents of the pitpercat can be expanded to include other familiar technoscientific entities – for example, the sofa and armchair.

In his final analysis, Michael broadens the co(a)gents. The interviewer, along with his tape recorder, the payment of five pounds sterling, his briefcase, and so on, evoke a co(a)gent such as Lancaster University (or even, conceivably, the university sector or academia-in-general). Conversely, the respondent's exclusive focus upon Burger King implicates a grander co(a)gent, namely Burger King. Michael suggests that the refusal to facilitate the interview indicates that the respondent is challenging the meaning of the interview situation. It has become a moment in which she redefines the interviewer's five pounds sterling: no longer does it signify remuneration for being an object of the academic gaze, it becomes the opportunity for enacting one's financial autonomy from the university sector, and one's economic and career alignment (or entanglement) with Burger King. As such one might say that this is a local, everyday moment in which such large-scale social entities (or co(a)gents) as a corporation (Burger King) and state sector (the University) come to be differentiated. In other words, in this moment, we see society in its 'structural' aspect being made.[5]

Michael then puts his three analyses together. For example, he notes that the monologue on Burger King is enabled by the co-presence of the pitpercat. This suggests that in the making of 'society' – in this case the reproduction of broad social divisions (between corporation and university) – nonhumans such as dogs and cats, and other everyday nonhumans such as technoscientific artefacts, play a constitutive role by virtue of enabling the respondent to enact herself as she does.

These analyses make the general point that technoscience in the form of mundane artefacts is pivotal to the process by which society and social data are produced. On the one hand, its co-presence allows certain performances that mediate 'broader' social entities and relations; on the other, through the work of the purifying sociologist, these have to be disciplined out of the equation, so to speak.

In the context of the critique of everyday life, Michael's analysis also suggests a number of ironies. It is possible to regard the role of nonhumans (most obviously the 'natural' entities 'pets', and various mundane technoscientific artefacts) as part of a complex nexus of heterogeneous orderings and disorderings. In the next chapter we will have reason to elaborate on such relations in terms of their spatialities – or, rather, metaphors of spatiality (e.g. prepositions). For the moment, however, we will focus on how such orderings and disorderings inflect with the concerns of critique. Drawing on De Certeau's analytic, the disordering of the intercorder can be seen as the break-up of a particular disciplinary spatiality into which the interviewee was to be drawn. A particular identity was being enacted (something like 'someone concerned with the local impact of the nuclear industry') and the intercorder (and the University) was to be a means of its 'surveillance'. The break-up of the intercorder might thus be thought of as a 'ruse', but a rather dramatic one: the exertions of the pitpercat become a form of resistance, a tactic. Perhaps extravagantly, we might say that what we witness in this process of disordering of the intercorder by the pitpercat, is an assertion of domestic heterogeneity in the face of the scholarly purification. At the same time, we might ask what this 'ruse' accomplishes in relation to the nexus of other associations in which the pitpercat and the intercorder (Burger King and University) are embroiled. As noted, this disordering serves in the ordering of institutional divides. However, the associations with Burger King also facilitate the 'pitpercat' (most obviously, financially), yet surveillance and discipline and purification (most obviously as a 'worker') are present in a different spatiality (the Burger King work space). As such the extrication from one moment of what we might call 'disciplinary purification'[6] (the research interview) and the disordering assertion of heterogeneity (the pitpercat) is partially mediated by another moment of 'disciplinary purification' (Burger King's corporate 'contract' with the 'worker', and the means of that contract's enforcement). Of course, the irony is that both purifications are possible only because of impurification: both sociological purification of the social subject and Burger King's purification of the 'worker' are deeply heterogeneous processes.

In this section, I have attempted to illustrate how 'society', in both 'micro' and 'macro' forms, is purified in everyday life characterized by mundane heterogeneity (in which technoscience is a key feature). The next section begins to look at how the communicative circulations of, and through, technoscience 'trace' society.

Technoscience 'traces' society

Michel Serres' account of the role of the circulation of quasi-objects in the making of society focuses on the way that these embody a certain meaning where 'meaning' is meant in the heterogeneous sense of 'order building' in which relations are established through signs or materials or both (see Akrich and Latour 1992). On this score, quasi-objects pass on messages, but not always reliably; indeed, sometimes, they are unpredictable (or under-determined) in so far as the meanings they mediate are still emergent. One might say that technoscientific artefacts of communication (ranging from Post-it® Notes through telephones to laptop computers), while they mediate messages, are also, to misuse McLuhan (1993), the message. While McLuhan famously, and controversially, emphasized the impact of media upon cognitive capacities (and bracketed their content), we wish to explore how they serve in the processes of ordering and disordering heterogeneous relations.

In what follows, I will consider two examples of technoscience that function as media of communication in order to address the ways that these 'trace' particular sorts of 'society': the Post-it® Note and information and communication technologies (ICTs). The Post-it® Note has been called the most significant development in office technology since the paper clip. As an everyday item it circulates through offices and households conveniently mediating little messages. However, its uses are at once tightly controlled and symbolically expansive. It mediates particular orderings and the possibility of ordering per se. So, while this seemingly trivial technoscientific artefact is typically regarded as mediating a depoliticized or abstract 'sociality', it can better be conceptualized as instrumental in the enactment of very particular relations of power. The virtual society is, it has been claimed, imminent – information and communication technologies will mediate new social relations at a distance: our communities will be altogether more dramatically distributed. Yet, in their material-social settings, ICTs generate a series of ironies in which older forms of society (face-to-face, local, 'unmediated') are re-entrenched.

In 1968, Spencer F. Silver, a chemist at the 3M corporation in the USA, developed a new form of adhesive. Comprising of highly stable microscopic acrylic spheres that attached tangentially to flat surfaces, this adhesive could attach and detach over and over again. However, a product that could make use of this low-tack adhesive was not forthcoming until Arthur L. Fry, in a moment of inspiration, realized it might be used for a restickable bookmark. After successfully trialling early versions of the Post-it® Note with 3M staff, the product was developed over the next five years and introduced across the USA in 1980. (see http://inventors.about.com/gi/dynamic/offsite.htm?site=http://web.mit.edu/invent/iow/frysilver.html, accessed 27 June 2003).

This, then, is the invention narrative that is associated with the Post-it® Note. It was a technoscientific innovation, which has since become a mundane artefact that operates more or less invisibly in the routines of everyday life. Indeed, nowadays, there are very many Post-it® Note products that cover the home, work and school, and include, most famously, varieties of notes and flags. There is even an electronic Post-it® Note for use virtually on the computer screen. Throughout the Post-it® Note website (http://www.3m.com/us/office/postit/), the overall tenor is of a convenient office supply product that quietly and efficiently does its job of communication. Indeed, communication per se is what the Post-it® Note signifies – society (at work, in the home, at school) is rendered in terms of abstracted communication or a contentless sociality. Obviously enough, such communication is about the drawing and directing of attention.[7] This is epitomized in links to a series of web pages relating Post-it® Note survival stories. For example, in 1996, a Post-it® Note, stuck to the nose of an aeroplane, survived the journey from Las Vegas to Minneapolis (a journey that entailed speeds of 500 miles per hour, and temperatures of -56 degrees Fahrenheit). Similarly, a Post-it® Note stuck to the back of a trailer survived a four-day, 3000-mile journey across America. Other illustrations concern Post-it® Notes that survived washing machines and dishwashers, or natural disasters such as hurricanes and floods. The messages might be faded, but the notes are intact, and in their original position (i.e. the adhesive has done its job). In these stories, the content of the message is irrelevant: it is the capacity of the Post-it® Note, in the face of extreme adversity, to maintain its capacity to pass on the message (whatever it is) that is celebrated; continued tackiness keeps the possibility of communication intact.

However, in the circumstances of everyday use, Post-it® Notes do not simply draw attention to themselves. They also point to those who place them, and, as such, serve in the subtle interplay of relations of power. Let us consider the example of Post-it® Flags. These are promoted in terms of their usefulness in drawing 'attention to critical items, for indexing and filing, or for colour coding' (http://www.3m.com/us/office/postit/products/prod_ft.html). Thus, we find these used in large documents, or collections of documents, to indicate subsections and to direct the reader to the most relevant or important part. There are even Post-it® Flags that incorporate arrows and messages such as 'Sign and date', 'Note' or 'Important'. These are represented as still more efficient ways for the everyday directing of attention. However, such directing of attention has very particular protocols attached to it. Superficially, Post-it® Flags conveniently direct attention to that text which is vital for the purpose at hand. However, the purpose at hand is not always transparent or unproblematic or uncontested. To have one's attention directed to one part of a document is to withdraw it from others that the reader might find more important or interesting. Thus, in some cases, administrators might direct attention as a service: the task at hand is held to be consensual (sign this; read this redrafted section) and attention is directed in the service of efficiency. However, in other cases, say if management directs attention and tries to represent this as a service (saving time; not distracting colleagues from more urgent tasks), it might also be read as a mode of control or illegitimate agenda setting. In other cases still, say when inspectors or auditors are reviewing documents, Post-it® Flags become highly inappropriate – under these circumstances, attention is expected to be free ranging. In each of these three examples, attention is related to particular forms of society. Put very simply these are, respectively, societies of: service providers and users (Post-it® Flags are a media of convenience); managers and colleagues (Post-it® Flags signal occasions for arguments over consultation or democracy); inspected and inspectors (Post-it® Flags are 'ruses' that aggravate the process of surveillance). Of course, we might expect complex variations and interdigitations of these. Nevertheless, our albeit superficial illustrations have the merit of exemplifying how a seemingly innocent device for communication mediates not abstracted communication or sociality per se, but contentful, complex societies. Needless to say, in the wake of Serres' insistence upon the mutual definition of society and artefact, while Post-it® Flags can mediate a range of societies, they are themselves reciprocally shaped by those societies.

If the preceding example evokes a world full of paper, our next illustration has often been associated with the idea of the paperless office (see Chapter 6). We now turn to the role of ICTs in everyday life, especially in relation to the possible emergence of something like the 'virtual society'. If Post-it® Notes as a means of communication physically carry their messages, ICTs act as channels or conduits for the two-way (interactive) movement of signs: the telephone and the computer are iconic in this respect. Of course with mobile phones and laptops, these artefacts move too, but they still serve the function of enabling communication at a (substantial) distance. Ironically, generalizing from Michael Bull's (2000) account of the personal stereo, such technologies can be seen as at once facilitating the sense of autonomy and control for users, while also typically tying users more tightly into the 'exigencies of late capitalism' as Poster (2002) phrases it. Indeed, such technologies might be seen from the perspective of a critique of everyday life as undermining 'reciprocal forms of recognition' (Bull 2000, p. 182). According to Bull's empirical analyses, people use personal stereos (and nowadays that would include MP3 players and the iPod in its various manifestations) as a means of isolating themselves from others in everyday life – as a means of entering more deeply into a private experiential realm in which control and autonomy might be exercised. It is this that Bull sees as a chronic, technologically mediated erosion of 'reciprocal forms of recognition'. Ironically, as Bull notes, personal stereo users are also accompanied 'by the mediated messages of the products of the "culture industry" ' (2000,

p. 160) – messages have predominantly taken the form of music, but also, increasingly, videos and podcasts (non-musical radio broadcasts converted for playback on MP3 players). However, with the increasing capacity to download other users' playlists of musical tracks, there arises, ironically, a minimal form of recognition of others-at-a-distance. With laptops and mobile phones, this irony is redoubled: to enter into a private world of the laptop or mobile phone is to come into communication – synchronous or asynchronous – with others-at-a-distance. That is to say, local reciprocal forms of recognition are seemingly replaced by forms of recognition conducted at a distance, across a 'virtual society'.

However, this account seems to me to be problematic because, while it deals with the experience of cutting oneself off and the erosion of reciprocal recognition, it fails to address the 'performativity' at the heart of such experience and recognition. Thus, 'cutting oneself off' locally can be conceived as a performance enabled by certain shared rules or expectations; accordingly, technology is used in a particular way that allows people to enact themselves as 'cut off', and, crucially, for others to recognize this cut-off-ness. Reciprocal recognition enters into this social (non)interaction by virtue of commonly accessible rules about how to do cut-off-ness more or less acceptably. As such, these rules (which, as the ethnomethodologists have taught us, can themselves be reflected upon and used in such social episodes) imply that the performance of cut-off-ness is mutually coordinated. Let us take a simple 'rule' (or expectation): that the personal stereo's volume should not be so high as to disturb others nearby. That a personal stereo is not played 'over-loudly' suggests that there is a recognition of a common expectation of ambient levels of noise. If it is turned up high, but the personal stereo user is not challenged, a decision has been made to suspend the rules, which implies something like a 'collusion' in the performance of 'cut-off-ness'. If there is a challenge, then this can be read as a plea for the unintended listener to be cut off from the noise of the personal stereo. Cut-off-ness is a collectively accomplished state, then.

In relation to such technologies as mobile phones and laptop computers, this apparent local cut-off-ness is at once 'necessitated' and 'replaced' by a mediated connectivity to others-at-a-distance. In terms of the local performances enabled by these technologies, there are a number of phenomena we can briefly note (also see below). These technologies – as bits of material culture – announce their user (not least through, say, the ringtone) as a particular type of person. Even the plainer or older models might be used to announce the user, as someone resistant to the 'gadgetry' and 'faddishness' associated with so many of the 'more fashionable' mobile phones. However, as a mediator of connectivity, the mobile phone can announce the user as busy, connected, popular, or potentially so – to carry a mobile in one's hand is to signal that one is in a constant state of expectancy of connection, to wear a

Bluetooth headset is to enact oneself as in a permanent state of electronic communicative readiness. In other words, such technologies serve as means to displaying one's 'extended sociality' even as one apparently 'withdraws', or cuts oneself off, from the 'local sociality' of the immediate setting, while nevertheless signifying furiously within, and affecting the character of, that 'local sociality'.

Poster (2002), in response to Bull's ambivalence about the personal stereo, suggests that, with the rise of ubiquitous computing in which 'all the power of information technology will be harnessed by and coupled with the body' (p. 756), we will be, potentially at least, party to a posthuman condition in which through these technologies a 'global space of cyborgs may emerge whose life task becomes precisely the exploration of identity, the discovery of genders, ethnicities, sexual identities, and personality types that may be enjoyed, experienced, but also transgressed' (p. 759). He regards this is as a more promising route to the realization of the 'everyday as festival' than Lefebvre's Marxist humanism. However, as the preceding analysis of local performativity in which technology is seen to render rather traditional forms of sociality suggests, such technologies are part of assemblages through which people perform themselves in pursuit of more or less typical (for the modern western, at least) social positionings or status (popular, important, fashionable, etc.). Moreover, as anyone who has recently travelled on a train in the UK will know, the performance of the local setting to one's interlocutor at 'the other end' of a mobile phone can likewise enact such claims to status. In other words, claims that these technologies will mediate such qualities as globality and virtuality need to be treated with due circumspection. On this score, we can turn to a recent critical review of such claims.

Following Steve Woolgar, these communication technologies can be collected with many others under the rubric of 'virtual society', which 'stands on behalf of concurrently emerging epithetized visions about technologically transformed futures' (Woolgar 2002, p. 3). Other terms that could have been used include, as Woolgar notes, information society, network society, global society. Virtual society is thus Woolgar's preferred term to capture those 'descriptions used to conjure futures consequent upon the effects of electronic technologies' (p. 3). Future worlds characterized in terms of such prefixes as tele-, cyber-, network, electronic (or e-), digital, remote, interactive, information and, of course, virtual, are in need of, Woolgar suggests, not only empirical scrutiny but also analytic scepticism. As such, he posits five rules for 'apprehending or making sense of the prospects for a virtual society' (p. 13). Rule 1 states that the 'uptake and use of new technologies depend crucially on local social context' (p. 14). Thus, for example, Wyatt, Thomas and Terranova (2002) show how people, under particular circumstances, actively do not use the internet. Rule 2 states that the 'fears and risks associated with new technologies are unevenly socially distributed' (p. 15). Thus,

Mason, Button, Lankshear and Coates (2002) show how the employees they interviewed, against expectation, did not necessarily fear the implementation of technologies that had a surveillance capacity because such technologies could be appropriated towards a number of useful ends. Contrariwise, managers could become more fearful of such technologies, whose surveillance and data storage capacities might be used in support of compensation claims for repetitive strain injuries. Rule 3 states that 'virtual technologies supplement rather than substitute for real activities' (p. 16). As Nettleton, Pleace, Burrows, Muncer and Loader (2002) document, computer-mediated communication enables 'a combination of ... shared experience and ... anonymity that contribute[s] to the quality of ... support. People using virtual systems of social support engaged in offering emotional support to others on the net, support that in "real" life we might imagine would be limited only to close friends' (p. 185). Rule 4 runs as follows: 'The more virtual the more real.' According to this rule, we would expect that 'the introduction and use of new "virtual" technologies can actually stimulate more of the corresponding "real" activity' (p. 17). Brown and Lightfoot's (2002) analysis of the introduction of network computing into organizations found that emails tended to supplement memos (indeed were treated as memos by other means) and face-to-face interaction (say, when social relations needed to be established in a team). However, emailing, in so far as it could be used to disguise dissembling, could also be seen to necessitate more face-to-face contact in that the latter could serve to establish trust in business dealings. Woolgar's final rule states: 'The more global, the more local.' As such, 'the very effort to escape local context, to promote one's transcendent (and/or virtual) identity, actually depends on specifically local ways of managing the technology' (p. 19). Cooper, Green, Murtagh and Harper's (2002) analysis of the local use of mobile phones brings us full circle to our account presented above. By contrast to the foregoing analysis, which focused, albeit peremptorily, on the performance of status and identity, Cooper *et al.* trace how in using the mobile phone on the train, gaze and gesture served to demarcate private and public domains (or, rather, served in their reconfiguration – what was private/public became much trickier to identify). Here, then, in order to enter into the 'global' as mediated by the mobile phone, actors enact various body techniques appropriate to, or at least recognizable within, the local setting in order to 'do privacy', even as the form and purpose of the communication comes to reinscribe what counts as 'public' or 'private' (see 2002, p. 295).

In summary, in the foregoing we have traced some of the ways in which new communication technologies, rather than propelling us into a realm of the virtual or transforming 'us' into members of a virtual society (even as this 'us' is a highly problematic category – e.g. Castells 2001), tie us further into the exigencies of the local, even as those exigencies are themselves reconfigured. However, as Woolgar elegantly shows, 'virtual society' is not simply a way of trying to describe the emergent world, it is also a means of making the emergent world; the many discourses associated with 'virtual society' can therefore be considered in terms of their performativity, as stakes in the production of society. It is because particular expectations attach to the virtual society that many of the findings presented in Woolgar's edited collection can appear 'counterintuitive'. The next section examines how such broad technoscientific characterizations of society might operate, not least in everyday life.

A Multiplicity of 'technoscientific societies'

In examining the confidence with which such terms as 'virtual society' are deployed, Woolgar identifies a number of constituencies (e.g. political spokespersons, the media, electronics industries, consultants, advertisers) that together, through interrelations and mutual support, evince certainty 'even in the face of setbacks to the envisioned scale of growth and development' (2002, p. 8) of the virtual society. However, rather than treat this as a set of discourses merely to be debunked, Woolgar insists that they are constitutive, not least in that they 'mobilize an array of different social constituencies' (2002, p. 8). In other words, we might ask what makes such representations of the more or less distant future so persuasive? This is a complex question and one that will be addressed in detail in Chapter 7, on time and temporality, where we will discuss the everyday ambivalences that such representations of the future entail.

Of more interest in the present context is that everyday life is pervaded by a multiplicity of such representations. Above, we saw how the term virtual society could function as shorthand for a variety of electronic communication technologies and systems. There are a number of such terms that aim to capture technoscientific developments in other domains. Thus, in relation to the new genetic technologies and the human genome project, one emerging term is the 'geneticization' of society, in which the gene becomes an icon that pervades popular culture and genetics is seen increasingly to explain a range of social and individual characteristics, not least those related to biomedical conditions (e.g. Nelkin and Lindee 1996; Lippmann 1998). As Conrad and Gabe (1999) put it in the introduction to their edited volume *Sociological Perspectives on the New Genetics*: 'We are at the dawn of a genetic age' (p. 1). Inevitably, there are other candidates for describing society apprehended through the new genetics: we have already encountered, in Chapter 3, the idea of posthumanity and 'Our Posthuman Future' (Fukuyama 2002). There is also the notion of 'biosociality' – a notion that aims to capture the ways in which 'nature is remodelled on culture understood as a practice. Nature will be known and remade through technique and will finally become artificial, just as culture becomes natural' (Rabinow 1996, p. 99). However, Hedgecoe (2001) reminds us that the empirical claims for increasing geneticization or the dawn of a genetic age might need to be treated cautiously. But more important is the 'work' this and related notions, and their associated socio-technical assemblages, might do. In an echo of Woolgar's argument, Lemke (2004) asks: 'What social consequences does the "geneticization of society" have, even if genome analysis and genetic diagnostics cannot redeem the high expectations we have of them?' For Lemke, geneticization needs to be thought of as a 'power strategy' through which the social is recoded as 'a space where the instruments of genetic diagnostics and genome analysis could both have an important role' (2004, p. 560). While the tradition (broadly Foucauldian) differs somewhat from that of Woolgar (broadly eth-nomethodological), the analytic ethos is similar.

Of course, there are also combinations of the techniques of biomedicine and information and communication technologies. We have already come across this in the form of bioinformatics. Another variant is embodied in biometrics – a biometric indicator is defined as 'any human physical or bio-logical feature that can be measured and used for the purpose of automated or semi-automated identification' (Institute for Prospective Technological Studies 2005, p. 11). This has precipitated concerns about a 'surveillance society'. The EU report on biometrics raises this as a widespread anxiety, and Richard Thomas, Britain's Information Commissioner, has warned 'that the country risks "sleepwalking into a surveillance society" because of govern-ment plans for identity cards' (*The Times*, 16 August 2004). At the time of writing (July 2005), there is a huge controversy over the proposed introduc-tion in the UK of identity cards containing biometric information, and while the controversy touches on a number of issues (not least cost), worries over a prospective 'surveillance society' loom large.

So, here we have three versions of a, more or less distant, 'technoscientific society': virtual, geneticized and surveillance. As we have seen, these are not wholly distinct in that they might well be sharing, or contributing to, the same sociotechnical assemblages. As accounts of (near-)future society, they 'perform' in various ways. For example, 'the geneticization of society' has been used, ironically, to reassert the importance of public health and health promotion policies (as opposed to, say, the apparent privilege afforded the development of individualized genetic therapies); and 'surveillance society' serves to provoke efforts to truncate what are seen to be the state's continuing efforts to gather private information on its citizens (as opposed to, say, the idea that identity cards afford better security). In other words, such rep-resentations are also about making society – persuading actors to see the world in a particular way, and to act upon, or shape action around, such a vision.

Now, within everyday life, there is a circulation of representations of such more or less disparate 'technoscientific societies'. Here, we have attended to virtual, geneticized and surveillance societies, but there are many others. Perhaps most obvious is what we might term 'environmental disaster society' – at the time of first writing (early July 2005) and around the time of the G8 meeting at Gleneagles, climate change and various technoscientific responses featured prominently in the UK media. For instance, in the UK, there is increasing controversy over the use of wind turbines (noise and unsightliness are represented as particularly troublesome), and nuclear energy seems to be becoming rehabilitated as a viable candidate for energy sources that are non-carbon emitting (also, nuclear fusion – as a 'clean source of unlimited energy' – has received more media attention by virtue of attracting major funding to build the 'International Thermonuclear Experimental Reactor' (*Guardian*, 28 June 2005)). If 'environmental disaster society' has signified mainly global warming in recent times, we can imagine that it will, if UK energy policy pursues the fission option, come to connote once again nuclear pollution.

The key point here is that there is a multiplicity of (shifting) representations of 'technoscience-in-society' in circulation, representations that are, arguably, woven into the fabric of everyday life, not least through the operations of the media. The range of 'technoscientifically tinged' societies that are more or less publicly available can be expanded to include society-affected-by-nanotechnology (or 'nanoculture'), network society, risk society, globalized society (or globalization), a society of somatic selves (where identity is realized through 'work upon the body'). To be sure, these versions of society might not always be explicitly articulated, and they will connect to, or merge with, other 'representations of society' (for example, celebrity culture, consumer culture, multicultural society, postmodern society). Nevertheless, these many representations (or perhaps representational fragments) of society comprise a multiplicity that suffuses everyday life.

Yet, how is such multiplicity 'dealt with'? Rather than an examination of the performativities of the various individual representations (e.g. geneticized, surveillance, virtual) of society, we might consider how this multiplicity, and the cross-circulations of 'technoscientific societies', feature in everyday life. In what follows, I will make some general and very tentative suggestions – suggestions that will need empirical investigation – as to the sorts of everyday reactions that might meet such multiplicity.

- Though this might be a slightly imprudent thing to say in a book on technoscience and everyday life, there is no necessity that these versions of society will make any impression on people's everyday grasp of society. For instance, approaching these societies in terms of their relation to risk, surveillance, geneticization, climate change might hardly feature at all where there are more pressing worries around, for instance, day-to-day economic survival (see Tulloch and Lupton 2003). In other words, we should not assume that these technoscientific society concerns play a central role in everyday life, either individually or cumulatively.

- However, rather than a passive indifference to this multiplicity, we might expect to find some people responding by active withdrawal: the multiplicity reflects a world out of control, or a world in decline (going to the dogs), or a world of knowledge and decision-making that belongs to others (governments, 'the experts') and from which one is effectively excluded. Ironically, then, such a proliferation of 'societies' leads to a simplification of society into a binary of 'us' and 'them'. As we saw in the previous chapter, governments have begun to try to overcome this bifurcation by encouraging public participation.

- The foregoing suggests a sort of 'negative' meta-principle through which to apprehend this multiplicity – a meta-principle marked by passivity or exasperation. Yet there can be more 'positive' meta-principles. For instance, the 'anti-globalization' movement might be said to embody such a proactive principle – a view that such societies are underpinned by globalizing capitalism that needs to be resisted and undermined. Conspiracy theories that attach these various societies to some overarching agent (e.g. the West, the US Government) might also inspire energetic interventions. Conversely, such societies might be seen as the sullied exemplars of 'pure' scientific principles (e.g. of the laws of physics, or the scientific ethos or 'meta-narrative – see Lyotard 1984).

- One response is that while there is an awareness of many of these societies, individuals or collectives might 'specialize' in one particular area – say one that is immediately relevant to their everyday circumstances (e.g. biomedicine, or surveillance). Moreover, they might see themselves as part of a broader collective, made up of groups each with its own specialism. In other words, the social world is imagined in terms of a sort of division of labour in which what is laboured over are differential versions of society that nevertheless 'fit together'.

- Finally, people might 'navigate' their way through this multiplicity by serially focusing on different issues as these become 'interesting' or 'fashionable' – or the objects of media or government attention. Unlike Mol's (2002) account of the 'body multiple' (see Chapter 3), there is no ready institutional impetus towards a 'resolution' (even in the unresolved sense that Mol develops) of these societies: cycling through is good enough.

The multiplicity of societies that emerges from the proliferating representations of technoscience hints at a multiplicity of everyday responses. Of course, this picture can be complicated in all manner of ways. These representations of 'technoscientific societies' do not simply coexist – they also compete as their various spokespersons promote them in one way or another. Moreover, what these societies are, or might be, is chronically contested. 'Geneticized society' has long exercised ethicists, sociologists, anthropologists, economists, politicians (e.g. Rose 2005). But it is also enacted through the accounts of scientific and biomedical professionals, who, as we might suspect, routinely disagree with one another. Even within a single specialist technoscientific domain, there is often a multiplicity of expert positions. This leads us on to consider the ways in which 'technoscience in itself' comprises a particular 'society' made up of disparate actors, entities and relations – a society whose complex representation circulates in various ways.

The society of technoscience

Once upon a time, science itself was held up as a model of the ideal democratic society (e.g. Merton 1973). Nowadays this vision is somewhat tarnished given the often bitter exchanges among scientists embroiled in public controversy, and the not uncommon accusation that scientists are obsessive, or hubristic, or opportunistic, or plain dishonest – from failing to subject themselves to proper peer review, preferring to announce 'results' or 'breakthroughs' directly to the media, through to outright fraud. The former practice of 'premature publicity' is often seen to be indicative of science's sometime need for a more prominent public profile, not least to attract further funding or more amenable regulatory circumstances. Here, science is no longer seen to be primarily concerned with the pursuit of knowledge.

Another source of people's acquaintance with science is, of course, fiction. There are numerous studies of the way that science and scientists are represented in fiction. A standard dichotomy in such genres is between the mad and the upstanding scientist – between scientists who would enslave or destroy society and those that heroically sacrifice themselves to save it (e.g. Haynes 1994). But, here, science is often reduced to 'scientists' and their various merits and/or pathologies. The everyday practices in laboratories through which scientific knowledge and technological innovations are developed are, typically, black-boxed – that is, the practices and processes whereby knowledge is generated remain unscrutinized. One of the most popular contemporary fictional representations of science-in-action is to be found in the recent spate of television police procedurals. Most relevant to present purposes is *Crime Scene Investigation* (or *CSI*) and its two spin-off series, *CSI: Miami* and *CSI: New York*, with their focus on the role of forensic science in solving crimes (usually murders). As we would expect to find in this genre, there is a preference for individualizing, pathologizing and moralizing explanations, over sociological ones, for 'deviant' behaviour. However, what interests us here is how the representation of scientific work in forensic science serves to render 'society' in relation to science. Typically, each episode of *CSI* has two concurrent investigations; also typically, the crime scene is deeply puzzling (which is presumably why the *Crime Scene* laboratory is needed). The *CSI* shows unfold as a dramatic, but patient, accumulation of evidence and the ongoing reinterpretation of the crime scene in light of that evidence until the perpetrator is identified. While there is material disorder in the outside world, within the *CSI* laboratory, all is materially ordered: laboratories are pristine, equipment works without breaking down, and the visualization of evidence, enhanced as it is by computer graphics, reveals the most obscure and intricate physical and physiological processes (e.g. internal bodily trauma). The vicissitudes of human bodies and machines are largely absent – the investigators routinely 'work the evidence' without the mistakes that are chronic in scientific laboratories (e.g. Amann and Knorr Cetina 1990). This is not to say that nothing goes wrong. There are plenty of problems but these are mainly social: a mainstay of the plotting are the troubles faced by individuals (e.g. psychological issues), or the tensions in social relations (e.g. career competitiveness), or unwanted institutional scrutiny or meddling (e.g. reorganization of shifts, or allegations of corruption). However, against these distractions, there is the refrain, common to all the characters in those moments where the plot demands that they play the rational foil to an irrational or distracted colleague, of letting the evidence 'speak' for itself. There are also regular exchanges in which new meanings from the evidence are deduced, and new hypotheses generated. But again the evidence 'speaks through' the investigator-scientists – they are its mouthpieces in an exchange in which individuals' brilliance as investigator-scientists lies in their very capacity to 'absent' themselves and let the evidence 'speak for itself'.[8]

In any case, much of the detail of the plot is driven by the work done to extricate society from the material evidential world, so that the material can reveal its inherent order. Unsurprisingly, then, we find in *CSI* that the sociality of science is merely a distraction from the 'seeing of evidence' – an impurification that must be removed so that the evidential order behind the material disorder is made visible. At the end of each episode this tension between material ordering and social disordering is reinstituted. Social mess reasserts itself in so far as each episode finishes with the hint of an ongoing institutional or interpersonal issue, something that will, in due course (i.e. in a subsequent episode), need to be extracted in the renewed pursuit of material order.

What we as viewers of *CSI* do not witness is that such evidence is 'seeable', in no small measure, due to the assumptions built into scientific procedures and techniques – assumptions that are themselves contestable, not least in the courts (a good example is the controversy over 'genetic fingerprinting'). This contestability – that any good lawyer should be able to pursue (e.g. Oteri, Weinberg and Pinales 1982) – is waylaid narratively, by the suspect's confession. Faced with seemingly incontrovertible forensic evidence, the suspect typically breaks down and confesses, thereby obviating any need to scrutinize in detail the evidence and the processes by which that evidence was garnered. The point here is that such popular representations portray science as a process of uncovering the order behind material mess in which the social is progressively excluded. In the expert protagonists' routine veneration of science, in their collective mediation of the 'voice' of the evidence, we are presented with a version of science that will proceed smoothly to the truth once social exigencies, and individual foibles, have been excised. The drama is as much about the struggles to expunge these contaminants from the scientific method, as about the identification of the real killer(s).

Instead, science can usefully be seen as an activity in which the material and the social are ordered together (e.g. Latour 1987a). A partial implication of this is that uncertainty and contingency is always at the core of scientific knowledge. Instead of the apparent certainties generated by *CSI* scientists (signalled not least in the way that suspects routinely break down and confess in the context of 'unimpeachable' evidence), the uncertainties that arise in science regarded as a heterogeneous process suggest that 'society' (with all its messiness) is a constitutive part of scientific knowledge: scientific knowledge is in this alternative view at once constructed and real (e.g. Latour 1999).

Concluding remarks

In this chapter, I have considered the complex and multiple emergence of 'society' in relation to technoscience and everyday life. Needless to say, my account is hardly exhaustive, though hopefully it is suggestive. We have encountered the impure purification of society through which the social sciences serve to divest the social of its nonhuman constituents. We have shown how technoscientific artefacts, which are claimed to mediate particular versions of society (e.g. abstracted sociality, virtual society), are much more complicated than these claims suggest. We have explored the pervasive multiplicity of 'technoscientific societies' and how this multiplicity might circulate through everyday life. We also touched upon the way that the social 'dimension' of technoscientific work (the society of technoscience) can be narratively constructed. In all this, paralleling our treatment of the body, everyday life seems to be characterized by two divergent apprehensions of society: as multiple and as singular. We will find that this disparity also characterizes the way that space and time feature 'in' and emerge 'out of' the circulations of technoscience in everyday life. If technoscience has been pivotal in realizing the linearities of singular 3-D physical space and clock time that pervade everyday life, it has, as we shall see, also been key to the proliferation of spatialities and temporalities.

Notes

1. This account of the role of the object differs somewhat from Knorr Cetina's (1997) influential analysis of sociality with objects. Knorr Cetina unravels the relations of knowledge experts to their objects of study in order to suggest that such relations are becoming generalized in late modern societies. Accordingly, there is 'an increased orientation toward objects as sources of the self, of relational intimacy, of shared subjectivity and social integration' (p. 23). In contrast, Serres' philosophical anthropology sees such objects always already constitutive of social relations, and though they proliferate over the course of history this does not seem to result in the sorts of qualitative change in social relations implied in Knorr Cetina's account.

2. These are the original *CSI* (set in Las Vegas), *CSI: Miami* and *CSI: NY* (New York). They have scored consistently highly in the US TV ratings, with the original *CSI* being number one, and the others in the top 20. The franchise is also proving very popular across Europe (see http://www.csiforum.no/news.asp?id=23&all=1).

3. This could be recast in terms of the process of governmentality (e.g. Dean 1999; Osborne and Rose 1999) in which social scientific techniques and the circulation of their results resource the way people come to comport themselves as social actors with particular properties or capacities (e.g. opinions).

4. Michael (2000a) uses the term co(a)gent to conflate the co-agents (pets and owner) and cogency (pets-and-owner).

5. To underscore, I am drawing on a microsocial reading of the 'social' here in which organizations are seen to be enacted in the local (e.g., Callon and Latour 1981; Knorr Cetina 1988; Law 1994).

6. 'Disciplinary purification' seems a useful term not least in that it connotes the operation of power (rendering particular orderings) and knowledge (the role of academic disciplines in this, notably sociology and economics in the present case).

7. Ironically, there are now super-sticky Post-it® Notes that 'are ideal for vertical and hard-to-stick-to surfaces, when you need your message to stay put and get noticed' (http://www.3m.com/us/office/postit/products/prod_notes_ss.html, accessed 28 June 2005).

8. Here, we are witness to an individualized version of what, in referring to scientists' typical representation of science, Traweek (1988) calls the 'culture of no culture'.

6 Technoscience and the enactment of everyday spatiality

Introduction

On 25 January 2005, there appeared in *The Times* a report about the irresponsible use of mobile phones by walkers to alert mountain rescue services. The opening paragraph read:

> The distress call from the mobile phone of a couple stranded up a mountain in the Lake District sounded serious. 'We are lost in the mist,' said an anxious voice. 'My wife is very frightened. Please come and find us.' As the message continued, the mountain rescue team listened incredulously. 'And could you send a helicopter?' asked the caller. 'We have a dinner date at 7pm which we really don't want to miss.'
> (http://www.timesonline.co.uk/article/0,,7-1454604_1,00.html, accessed 24 September 2005)

From 2004, the Wales Tourist Board ran an advertizing campaign promoting Wales as 'The Big Country'. In a series of four posters, Wales is contrasted to modern, urban everyday life. For example, in one poster, the number of castles in Wales (461) is juxtaposed to the number of Starbucks (6) as if in a score line. In another, headlining a spectacular photograph of Snowdonia is the mock category 'area of outstandingly bad mobile reception' – a parody of the UK Countryside Agency's official designation 'Area of Outstanding Natural Beauty' (AONB).

The text continues: 'Travellers riding up the Snowdonia mountain railway may experience communication problems. Your boss can't reach you. Even dogged telesales reps struggle. Damn those impenetrable mountain passes. Damn them' (see http://icwales.icnetwork.co.uk/yourwales/tourism/tm_objectid=14980940&method=full&siteid=50082-name_page.html, accessed 24 September 2005).

In these two examples, we see two traditional spaces – the city and the countryside – related to one another in complex ways, ways partly mediated by the more or less everyday technology of the mobile phone. In the *Times* piece, there is indignation at the use of the mobile phone to summon mountain rescue, which has been reduced from an emergency to a taxi service (or transforming the rescue personnel, as the article headline states, 'from life-savers to nannies'). We can read this indignation as informed by an amalgam of concerns about the nature of nature's space. If it were once a place of physical danger or the sublime (e.g. Thomas 1984), it has now become thoroughly domesticated – suffused with urban everyday life in the form of seemingly permanent contactability. Rephrasing this, we can follow Amin and Thrift (2002) and note that the city has marked the rural with its 'foot-print'. Indeed, it is not just the case that there 'is' accessibility, but that it is taken so much for granted that visitors' comportment in the countryside can be judged in terms of it. More particularly, the use of accessibility becomes 'irresponsible' or 'improper' because of what is seen to be an illegitimate transposition of modes of behaviour common to city life into the country-side. This criticism concerns not simply unreasonable phone use, but, because of the very possibility of contact, a tendency towards 'unpreparedness', or lack of respect for the risks posed by the countryside. After all, this article is part of a long line of reports admonishing walkers who have failed to prepare themselves for the dangers of nature (most obviously, the changeability of weather conditions) and relied on the mobile phone to access the rescue services. Of course, there are also reports celebrating such rescues, but these feature walkers or climbers who have come 'prepared' – that is, walkers who are already 'moral' in their comportment in the countryside.

In the complex of admonition and celebration, the countryside is at once contrasted against the city (as it has long been – see Williams 1973) and seen to be 'domesticated' by it. The rural is at once 'porous' to the urban (Amin and Thrift 2002) – not least through the reach of communication systems – and resistant to it not least by virtue of its unpredictability. Such re-presentations of 'near-wilderness' or 'potential wilderness' spaces, are marked, following Callon and Law (2004), by both presence and absence: the material-semiotic presence of the countryside and absence of the city, and vice versa.

Put another way, the connectedness (and the lengths and densities em-bodied in that connectedness) that characterizes the urban is ambiguously related to the rural. Being 'in nature' in this context is typically marked by an, albeit highly elastic, level of self-sufficiency, that is, of being able to look after oneself.[1] Such self-sufficiency is about the temporary or, rather, the 'tactical' shedding of everyday connections ('tactical' in the sense that to do self-sufficiency, shedding needs to be planned and ongoingly reviewed). The con-tinual connectedness afforded by the mobile phone must thus be monitored, and drawn only upon in those extreme moments where self-sufficiency fails – moments that, as the *Times* article illustrates, must be constantly policed. However, the identification of such 'failure' of self-sufficiency entails a complex judgement that ranges over such considerations as, as we have seen, the unforeseeability and severity of conditions and the level of walkers' preparedness, but also, for instance, the desire to attract visitors from the cities, and the role that mobile phones, by virtue of promising a certain degree of security, serve in that attraction. In other words, it is not simply that mobile phones mediate a connectedness that muddies the romance of the countryside, but are partly constitutive of a version of the countryside in which the tourist gaze (notwithstanding all its ironization) is welcomed (Urry 1990). The mobile phone is thus 'fluid' (Mol and Law 1994), changing its shape as it moves from 'context' to 'context'. Or, rather, its multiplicity of meaning and uses reconfigures that 'context' – articulates with the latent meanings of the urban and countryside to generate contrasting but co-present spatialities (see below).

However, there are further contrary aspects to this connectivity. It is now possible to enhance mobile phones (via such additions as Nokia's Xpress-on^{TM} GPS shell) to enable connection to global positioning systems. Given that GPS can inform users of their position and nearby geographical features, they can facilitate 'self-rescue', as it were; self-sufficiency is tied still further to such broader circulations. Indeed, it is tempting to speculate on how long it will be before the presence of GPS equipment becomes embroiled in moral judgements about the responsibilities of walkers.

The interplay between the urban and the rural is also evoked, ironically this time, in the Wales Tourist Board advertisement. Putative everyday contactability in the countryside is subverted by those damned 'impenetrable mountain passes'. However, let us recall that the problems of mobile phone use arise on the Snowdonia mountain railway. It would seem that Snowdonia's space is already partly domesticated – relative convenience and comfort characterize this particular train journey. Paradoxically, it is the incursions of bosses and dogged telesales reps that evoke the wild – wild in the sense of entailing a degree unpredictability. Further, supplementing the account of mobile phone use, presented in Chapter 5, such calls are not simply a means to performing 'importance' (the need to be constantly contactable), but are seen potentially to undermine being 'elsewhere' from the mundane, to undo the specialness of 'being in the mountains'. The irony is, of course, that this undermining is itself undermined by those damned 'impenetrable mountain passes'. Here, then, urban and rural spaces fold into each other in ways that are at once mutually sustaining and co-sabotaging.

Let me list a number of themes that can be derived from these everyday stories – themes that will be used to structure the rest of the chapter. First, as we have seen, these contrasting spaces of the rural and the urban are not givens: they are produced in a series of discourses and practices that encompass, for example, physical risk, technological capacities and moralized comportment in varying proportions. That is to say, such spaces have a history and it will be important briefly to sketch how we might think through some of the ways in which everyday space has changed over time. Second, and relatedly, we will need to consider how everyday space is sensed and mapped with the aid of, and in apparent contrast to, technoscience. Being frightened on a mountainside is intimately woven with the representations of the countryside – through Ordnance Survey maps, through photographs of panoramic vistas, through GPS. If our everyday feel for space is mediated in part by technologies that serve in the representation of that space then, arguably, there is an element of reduction – or disaggregation – of the elements of space that can be counterposed to seemingly more 'holistic' local sensings and understandings. On this score, taking such technologically mediated representations as performative, we explore how persons in such spaces come to find themselves 'performed'. Third, there is the complex issue of 'belongingness' or 'fittingness'; that is to say, the issue of what 'goes' with what in the process of spatialization. The mobile phone-carrying walkers might be seen, to borrow from Mary Douglas (1966) as 'matter out of place', 'dirtying' the countryside with their unwarrantable behaviour. However, this suggests that the 'place' of the countryside is somehow fixed, whereas it might be suggested that their actions remake that space, comprising not so much a contamination as a reconfiguration. In these 'productions of space', to borrow from Lefebvre (1974), we will touch upon such concerns as comfort, surveillance, loss, risk, domesticity and wilderness, hybridity and borderlands. Central to such concerns is the issue of how to attend to the various ways that entities 'come together' with other entities in the making of space. Specifically, we explore some of the sorts of prepositions that might allow us to grasp – or at least, address – the relations into which entities enter, and out of which they emerge, and, in the process, to re-examine the role of technoscience in everyday spatialization.

Theorizing spatialities

In this section, we briefly consider how space has been thought of not as an 'empty' volume in which events or entities or relations are contained, but as a historically contingent category. In what follows, we trace some of the ways in which 'space' has emerged in relation to everyday life and technoscience, from the 'opening up' of new spaces for everyday occupation to the conceptualization of the everyday, local space in terms of the processes of globalization. In pursuing this idea of the history of space, we also begin to identify the shifting parameters along which we might account for the changing character of spaces.

As is well known, Henri Lefebvre (1974) developed a linear history of space in which particular spatial configurations were associated with particular epochs, especially different modes of production. Thus, for example, in the capitalist period, space undergoes increasing fragmentation, instanced in the production of divergent areas – ghettos that each accommodate only members from particular social and economic strata. In the arrangement of ghettos were reflected the social and economic hierarchies of the epoch. Shields (1999), however, argues that for all of Lefebvre's brilliance at periodizing space, he has nevertheless imposed a sequence of epochs 'that directs attention away from struggles in everyday life to grand themes in the economic and political structure of a time' (p. 172). Of course, this does not negate the value of such historicization of space – it simply means that the differing versions of space need to be studied in their concrete specificities. The nuance and texture of particular spaces cannot be accounted for in terms of the dominant spatialities of particular epochs, however much (or dialectically) those epochs accumulate and interact within and with any such space. Having noted this, it is useful to try and unravel some of the parameters by which everyday space might be understood, even if, inevitably, such parameters are of their time and place.

Space, according to Stephen Kern (1983), underwent some major changes over the course of the early part of the twentieth century, not least in terms of its 'habitability'. For instance, spaces that were deemed 'empty' because darkness made them inhospitable became 'fillable' with the spread of electrification and illumination. Kern sees this as a general trend in which empty space was rendered positive or full – spaces that ranged from interior, psychic spaces to the architectural spaces between walls. However, we might also add (echoing Kern's analysis of the emergent perspectivist views on the contents of space at the turn of the twentieth century) that 'emptiness' is perspectival: spaces are not empty per se, but empty of, or resistant to being filled by, certain contents (e.g. 'civilized' persons). 'Filling' thus entailed reordering space: observing and removing certain elements, introducing others. Illuminating the streets of cities thus exposes certain relations that become the object of surveillance and ordering,[2] and once reordered allow for 'appropriate' filling.

In light of this, we might identify a hybrid parameter that encompasses emptiness/fillability, darkness/illumination, safe/unsafe. At the same time, as such writers as Simmel and Benjamin noted, urban space is characterized by excess: there is an overabundance of stimulation in the modern city. Space is marked – indeed, 'produced' – not so much by the possibility of 'filling', but by chronic 'overfilling'. That is to say, contemporary spatiality is typified by multiplicity. As such, everyday urban life necessitates, in one way or another, the sensorial management of material and semiotic plenty. At the beginning of the twenty-first century, as the work of commentators like Urry (2003) and Amin and Thrift (2002) suggests, and as the opening example of mobile phones in the countryside illustrates, it is perhaps more useful to say that 'space' is an analytic problem by virtue, not of being 'enclosable' (say through illumination or surveillance), but of being *endlessly disclosable*. Nowadays, spaces might better be defined in terms of the multiplicity disclosures that they can be made to yield.[3]

This multiplicity is partly the upshot of the number and range of relations and connections that 'come together' in (order to produce) a given space. This issue, obviously enough, maps onto attempts to theorize the 'interactions' between the global and local. Collected together 'in' everyday space are signs and materials whose origins lie in far distant parts of the globe. Indeed, such collections – say on the aisles of a supermarket – juxtapose products from all over the world and in such juxtapositioning are said to 'compress' space (and indeed time – see Harvey 1989) in that they signify the 'overcoming' of distance. For John Law (2004), however, in so far as these juxtapositions entail the interaction of the global and local, this is not best studied by tracing the 'global' elements that converge to make the local – what he calls 'looking up', or the study of romantic complexity. Rather, (baroque) complexity is seen to be contained in the detail of the local: to access it, we need to 'look down'. In such detail we find that the global is already present. As Law puts it, 'the global is situated, specific and materially constructed in the practices that make each specificity ... it is specific to each location, and if it is bigger or smaller, it is because it is *made* bigger or smaller at this site or that' (2004, p. 24). If we return to the examples of the mobile phones in the local sites of Cumbria or Snowdonia, we see that the 'global' evidenced in the reach of the mobile phone network is made, in the two cases, respectively larger and smaller (also see Franklin, Lury and Stacey 2000).

Still, part of making the local is that signs, bodies, materials, forces, energies congregate. These elements can be seen to come from the 'global' – to have travelled distances enabled by various networks and conduits that straddle the globe. But, this picture of the 'shortening' or overcoming of distance that is said to be characteristic of globalization (e.g. Giddens 1990) rests on a view of distance as an abstract quantity that is lessened or increased by, for example, such technoscientific assemblages as the internet or transportation systems. In contrast, distance can be treated 'qualitatively'. We might see it as comprised of many local settings linking up to make a conduit. The articulation, calibration and management of such locals are themselves local matters, as indeed is the representation of a 'conduit' and the related sense of a lessening of distance (i.e. globalization). In relation to Law's baroque perspective, this account of the global suggests a consecutive downward gazing – seeking the global in one local after another.

However, this account suggests a movement across discrete locals whereas, as Mol and Law (1994) and Urry (2003) have suggested, it might be better to consider such spatiality in terms of fluidity. As noted above, Mol and Law have focused on the way that the movement of entities across locals changes those entities (their classic example is that of 'anaemia'). However, local spaces change too as entities such as anaemia or the mobile phone enter them. In other words, it might behove us to think in terms of a 'mutual warping' between an 'entity' and 'its space': a target entity such as a mobile phone imports aspects of its 'previous spaces' into new spaces, thus more or less transforming the latter; conversely, new spaces affect the composition and constitution of the target entity.[4] It is clear that in such a formulation the contrast between 'entity' and 'space' becomes thoroughly blurred. To be 'looking down' at an entity is thus to disclose the multiplicity of spaces 'contained' within it. In relation to technoscience and everyday life, to 'look down' is to examine the multiple spaces entailed by an object like a mobile phone whose 'spaces' include, as we have seen, telecommunications, rescue services, romanticism, the cultural borders between countryside and city, the distribution of responsibilities. It is, indeed, to engage in what above we have called the 'endless disclosability' of contemporary space.

In the next section, we examine this view of everyday space in relation to the 'sensing' of space – in particular, the ways that everyday life is complexly mediated by both mundane technoscientific artefacts and technoscientifically abstracted representations of space.

'Sensing' space

On one level, everyday space needs to be understood as part of an ongoing process in which the perception of the world (and that includes space and its stuff) is orientated to particular ongoing actions and existing and emergent goals. In such an 'ecological' account of perception (e.g. Gibson 1979; Ingold 1993), 'space' is regarded as highly specific: it emerges along with, and relative to, the actions of the actors within it. Perception on this view cannot be separated from actions in the world. Now, such everyday perceptions (which are 'pre-analytic', as Ingold has argued) are enacted through a complex of mundane technological artefacts. Bodies, in engaging with the stuff of spaces, do so in a range of ways that entail such bits of everyday technology as clothing, shoes, spectacles, contact lenses, hearing aids, and so on and so forth. Our perception of such spaces is thus partly mediated through the use of such mundane artefacts.[5]

However, the use of such artefacts is also performative. They serve in the making of relations with other actors – human and nonhuman. Let us take the example of sunglasses. Carter and Michael (2004) note that sunglasses enable people to see in bright sunshine, lessening the need to engage in such actions as shielding the eyes with the hand, or keeping to the shade, or ensuring that the brim of one's hat casts a shadow over one's eyes. In the process, they affect the perception of a scene: space is aestheticized in so far as it is framed in a particular way and seen through the palate of colours filtered by the lenses. Of course, sunglasses also aestheticize their wearer in so far as they affect how the wearer comes to be seen. The brand or style of sunglasses, how they are worn on the face, how they relate to whatever else one is wearing – all these function in the performance of a particular self, a performance that goes on to affect the space in which the wearer moves. Carter and Michael (2004) list a number of cultural indices for how the wearing of sunglasses might be read.[6] For example, they mention how mirror glasses – say, through their filmic associations with maverick police officers – can generate not only inscrutability but also a sense of threat, or at least surveillance. Inevitably, such making of space in the act of its 'perception' – the 'mutual warping' of space and artefact – is further complicated by the proliferation of technologies available to us: cameras, camcorders, mobile phones, and so on.

We must not neglect the fact that such spatializing performativities are coordinated in everyday interactions. That is to say, people might use such artefacts in making particular spaces, but they are themselves operating in spaces that are being simultaneously made by others. In other words, we are also performed by others' uses of technoscientific artefacts. The use of personal stereos, as Bull (2000) shows, affects how listeners comport themselves in relation to others, but also how others orientate themselves to listeners. In abstracting the key insights of Goffman's corpus, Manning (1992) notes that for Goffman there are typical 'assumptions' about everyday social interaction. Relevant to the present discussion of spatiality are such 'norms' that, on the one hand, make's oneself accessible to others ('availability' – e.g. there is an expectation that one will respond when asked the time) and, on the other, one withdraws one's attention from others ('civil inattention' – e.g. there is an expectation that one will not subject others to one's unrelenting gaze in a lift). These 'rules' concern the enactments of mutual spatialities, and are affected by expectations around the meanings of technoscientific artefacts. As we have seen, the personal stereo (and the doing of 'being lost in the music') is an indication of being 'unavailable', of 'removing' oneself from common space – or cocooning. Part and parcel of this is that the listener performs others as outside of the cocoon. Needless to say, the listener is being tangled in the technoscientifically mediated performativities of others. Here, then, we have a picture in which spatiality comprises the complex and shifting borders of personal spaces, and the warrantable possibility of entering or protecting such spaces. While we have focused upon the role of the personal stereo and mobile phone in this, there are of course many other mundane technologies that are instrumental in this spatialization: the position of carrier bags and briefcases, the use of books and newspapers as barriers to gaze, the ostentation of anxious glances at watches or the dramatic removal of spectacles. Space, here, emerges from such mutual performativities (or warpings) enacted by persons-and-their-artefacts interacting with persons-and-their-artefacts.

If the focus thus far has primarily been on the visual modality, in general, the experience of space typically entails a range of sensation: sights, sounds, smells, touch, proprioceptors – all contribute to the sense of spatiality. Indeed, our shifting relation to particular spaces is shaped by cumulative and complex experiences – experiences that are partly mediated by mundane technologies. The richness of these experiences – or, to put it another way, the density of these spaces – is affected by all manner of specificities: particular modes of transportation, narratives about a given place, the circulation of divergent representations of 'generic' places (e.g. the 'sublime' mountains, the degenerating earth), different forms of 'gaze' (or, more broadly, forms of sensing), the presence or absence of various companions, and so on. As we shall see in a later section, such local spatialities have complex relations with 'global' dynamics.

Now, as is well known, prolific sensorial variegation has been treated as a 'problem', most famously by Simmel (e.g. 1979), who posited the 'blasé attitude' as a means of coping with this. For Benjamin, it was the *flâneur* and his (*sic*) reflexive engagement with the stimulatory multiplicity of the city that allowed it to be grasped (however 'poetic' such a grasp might be; see Ash and Thrift (2002) for a thoroughgoing critique of this analytic character). Alongside the variegation of such experiences of space, there is also the 'reduction' wrought by technoscientific representations. This reduction – what we might think of as the systematized loss of detail – is what allows technoscience to 'penetrate' and analyse its object (or space) of study (see Lynch and Woolgar 1990) and thus to reveal something profound, essential or generalizable about it. Indeed, the systematization of representational space (crucially in the form of Cartesian space), enables the accumulation and combination of huge and divergent data sets into such comparatively simple representations as graphs and tables; it is this property of scientific representations that, according to Latour (1990), lends them their peculiar potency. Of course, such insistent readability (or immutability in Latour's terminology) rests on particular skills, a certain credulity and local processes and procedures of negotiation. Thus, as Traweek (1988) shows, experimentalists in particle physics subject the representations of their experimentalist colleagues to much fiercer readings than do theorists. The point is that technoscientific representations of innumerable types reveal something 'profound' (or indeed practical) through simplification of one sort or another. However, this simplification can in principle always be complicated, and the form of complication reflects local contingencies.

So, technoscience has allowed us to penetrate the most obscure spaces: from the subatomic to the cosmic level, technoscience depicts 'spaces' that lie beyond mundane experience. Crosscutting everyday apprehensions of space is a sense that such spaces are shot through with various 'unsensibles'. At the extremes, space is the domain of quantum and cosmic processes; in-between everyday space is cohabited by chemicals and rays, bugs and forces, whose existence is putatively 'revealed' (and sometimes the upshot of) the workings of technoscience. For example, we mentioned above how spaces can be penetrated and visualized through GPS technology, but there is a plethora of other technoscientific 'visualizations'.[7] Such technologies 'reveal' aspects of spaces that would 'normally' be invisible to human senses – or, rather, they constitute spaces in terms of these detectabilities. For instance, everyday space is infused with electromagnetic waves from a variety of sources (e.g. overhead power cables, communications systems). These have been regarded by some as posing, or potentially posing, risks. Mobile phones have been linked with brain tumours, though recent research has again suggested that there is no increased risk of cancer (http://news.bbc.co.uk/2/hi/health/ 4432755.stm_, accessed 3 October 2005). To waylay even the dim possibility of such risks, hands-free sets are routinely recommended to minimize risks in the absence of long-term studies. In this example, space is 'warped' by technoscience into differential domains of doing and not-doing. That is to say, how we 'perform' space inflects with how, in light of more or less trusted expert visualizations of space, space 'performs' us.

We can expand on this example. For those who identify chronic electromagnetic pollution as a source of risk, the Teslar watch has been developed, which, with the use of 'scalar energy', 'magnifies the strength of the biofield thus protecting the body from destructive electromagnetism' (see Bioenergyfields Foundation, http://www.bioenergyfields.org/index.asp?SecId= 5&SubSecid=26, accessed 3 October 2005). Not unexpectedly, the Teslar watch has been denounced as quackery or expensive fashion masquerading as prophylaxis. The broader point is that space can be visualized or mapped not only in terms of this or that mode of detection (say, of electromagnetic radiation), but also in terms of uncertainty and contestation. Everyday spaces in this instance, through the evident contestability of the technoscience, are thus 'probabilistic' in the sense that space might be inundated not only with risky invisible physical properties, but also with risky social relations. In such hybrid spatial contexts, issues of trust, identification and differentiation loom large: one 'sees' the particular character of such spaces – or rather one aligns oneself with such spatializations – because of the sorts of trust one invests in, or identifications one makes with, particular 'expert spokespersons' for such spaces.

Another example of this juxtaposition of expert and experiential spatializations can be found in Macnaghten's (2003) analysis of the contrast between specialist (whether NGO, business or governmental) accounts of environmentally threatened 'global nature' and people's everyday experience of nature. His respondents, while certainly aware of the iconic representations of 'global nature', could best relate to images of the environment when they were connected to the 'personal realm of everyday life'. Direct connections to the local and personal environment tended to 'hit home' and 'matter more' (2003, p. 74).[8] This local spatialization of nature was further underlined by participants' accounts of their encounters with local environment in which nature served as a 'source of pleasure and transcendence from the burdens and stresses of everyday life' or comprised a 'setting for maintaining social ties and bonds', or constituted a range of problems on which people had to act if they were to fulfil their perceived 'responsibilities as mothers and parents' (2003, p. 77). In Macnaghten's account, there is a tendency, ironically, to naturalize these engagements with, and relations to, the local and the everyday. If the status and trust associated with 'global nature' appear tentative, the apparent 'authenticities' offered by 'local nature' need to be treated with like circumspection. After all, as we have noted, the everyday – including the everyday of local nature – is itself a construct whose character is much debated. Over and above this, the specific everydays that Macnaghten contrasts to global nature themselves emerge from a range of sociotechnical assemblages. Most obviously, as Macnaghten and Urry (1998), among many others, have shown, the idea of local nature as a 'source of pleasure and transcendence from the burdens and stresses of everyday life' reflects generalized romantic attitudes in the western perception of the environment related in complex and contradictory ways to industrialization and the idea of the countryside as refuge. Further, the countryside has in the past signified, and in many ways continues to signify, hardship and drudgery, resource and utility. The point is that the version of the local nature of everyday life that comes to be juxtaposed to distanciated global nature is itself no less 'global' – the partial upshot of circulations of particular versions of nature (and of the appropriate comportment towards that local nature). Or, to put it in Law's (2004) terminology, to look down into these everyday activities is to see the 'global'.

Now, as hinted, the above account can be couched in terms of the patterning of trust in relation to depictions of global nature and experience of local nature. But trust is not always consciously held, as various commentators (e.g. Giddens 1991; Stzompka 1999) have noted: more often than not trust is something that is presupposed, unarticulated. In the everyday use of a map, as with other mundane artefacts, we do not routinely reflect upon, let alone problematize, its provenance and trustworthiness. Indeed, the experience of, and engagement with, everyday practical space often incorporates such technoscientifically mediated representations of space as the map. That is to say, local spatializations entail such artefacts: maps, compasses, street signs, mobile phones, and so on, are an integral part of 'moving through' both the city and countryside. As we saw in Chapter 3, in relation to the issue of comfort (e.g. around ambient temperatures), our sensing of what is comfortable is rarely called into question – we have become used to enacting ourselves in such 'comfortable' spaces (even if that means making little adjustments as to how we act). Nevertheless, such automaticity is often challenged by the vicissitudes of spaces. Maps can always be rendered 'inaccurate' by virtue of local changes (traffic accidents, landslides). In these contexts, they are, obviously enough, also useful in finding alternative routes. But in such re-navigations, ongoing, iterative judgements based on consecutive local conditions will also be fundamental. In order to make the map workable, spatial detail must be added – for example, assessments of weather conditions or evaluations of the state of the route (e.g. road surfaces or congestion). Here, then, we see an interplay between formalized, technoscientifically reduced space (the map) and local experiential space (we shall echo this 'distinction' – it is almost tempting to say 'dialectic' – in the next chapter when we contrast proper and clock times).

While the two 'registers' often 'supplement' one another, they can also clash. Where particular spaces are experienced multisensorially, collectively and historically, technoscientific representations can variously warrant or query the uniqueness of such spaces. For instance, in Weldon's (1998) study of the public inquiry into Manchester Airport's second runway, it was clear that local residents valued the threatened area because of a nexus of apprehensions that touched on daily routine, biographical detail, seasonal change, aesthetic pleasures and intergenerational responsibilities. By contrast, expert spatializations disaggregated such complexities into noise contours, or parts per million, or population levels of the endangered great crested newt and corridors between newt habitat sites. In the space of a public inquiry, particular forms of representation typically take priority: the reduced representations of technoscience are seen to have greater credibility than the impressionistic – and sometimes literally poetic – accounts of the 'community'.[9] However, as hinted above, these registers can supplement one another. The everyday experience of particular spaces can be said to be 'deepened' – in the sense of entailing more connections (see Latour 2004a) – by the addition of apprehensions furnished by expert analyses of space (e.g. technical knowledge about great crested newt habitats).[10]

This section has briefly explored how representations furnished by technoscience inflect with everyday apprehensions. In the process, we have noted how these inflections are deeply ambiguous: quotidian space can be both 'enriched' and 'diluted' by such representations. As such, everyday contested spaces are marked by contrasting performativities: people enact themselves as an integral part of particular spaces, and enact spaces as an integral part of themselves. Technoscientific representations can serve in both the tightening and loosening of such couplings. At the same time, such tightenings and loosenings are deeply moral and political, and touch on matters of trust, risk, identity, community.

The next section considers further the complex emergence of everyday space under the broad rubric of 'what goes with what'.

What goes with what?

In this section, we will consider how space is constituted by combinations of stuff – that is, relations between various entities (including technoscientific artefacts) that generate particular everyday spatializations that incorporate both the 'local' and the 'global'. Moreover, we explore relations that, though obvious in everyday life, have been empirically and analytically neglected. Despite their relative obscurity, these relations can nevertheless usefully illuminate some of the ways in which spatializations are produced and countered. The first concerns the domestic 'sphere', in particular the role that mess and clutter, and the various discourses and practices associated with clutter and decluttering, play in the multiple and serial spatialization of everyday life. The second concerns the rural–urban 'borderland' especially as embodied in the figure of 'roadkill', and traces some of the complex spatializations wrought through the respective movements of animals and motor vehicles, and their 'intersections'. Both examples point towards a rethinking of spatiality in terms of a 'philosophy of prepositions' to be taken up in the conclusion.

The ordering of clutter

To think of the spaces of everyday life is often to imagine movements that are more or less lubricated or linear or, alternatively, more or less hampered or diverted. The conduit of one's daily commute is blocked because of problems with transport systems; the normal trajectories of a shopping trip are deflected because of a faulty credit card; easy movement through the home is frustrated by sheer clutter. The point is that, common sensibly, to 'traverse' a space, that space needs to be kept 'clear'. Of course keeping things 'clear' is hardly a passive exercise (and the term 'traverse' obscures the process of mutual warping out of which both traversee and space emerge). To state the obvious, keeping the trains and buses to time, ensuring one's credit card does what it is meant to do, are matters of great sociotechnical logistical complexity.

By comparison, keeping the home tidy seems a simple and menial task. Needless to say, the comparison does the labour involved in domestic space clearing an enormous injustice, not least because such labour and its spatializations are highly gendered. On the one hand, the many domestic devices that have supposedly made women's lives easier have, paradoxically, through a range of factors, quite possibly made domestic life more arduous. The easy movement 'through' domestic space is hindered for women by the many and increasingly more involved tasks they must attend to just to keep that space 'clear' (see Cowan 1985; Bittman, Rice and Wajcman 2004; Gershuny 2004).

Moreover, as we would expect, 'looking down' into such domestic devices reveals complex spatialities. The refrigerator and freezer, for instance, embody a series of spatialities that include a local 'climate in a box', the shapings and reshapings associated with changes in kitchen design (and relations to other appliances), the material and semiotic circulations that enable cooled and frozen products ('a new economy of cold') to arrive at, and be housed in, these appliances (Shove and Southerton 2000; Watkins 2003). However, such appliances also entail particular gendered corporeal spatializations: as Silva (2000) has argued, the use of such domestic appliances is regarded as 'artless' because the appliances themselves are marketed as 'intelligent' (as opposed to the use of masculine gendered technologies – typically IT or DIY – where skill resides in the man).

Now, such domestic skills have, to state the obvious, long been subject to moralization – the good wife and mother ensures that her good home is tidy and clean. In the UK, tidiness, decluttering and storage have nowadays increasingly become a focus of expert and media attention. As Cwerner and Metcalfe (2003) document, there are several "gurus" of the storage revolution who aim to unburden people of a domestic life cluttered with too much, often unneeded, stuff that gets in the way of an 'efficient' or 'rational' life. For example, a quick web search throws up a list of several 'professional organizers' for England. As one entry for the 'no more clutter' service puts it, 'no more clutter is a professional decluttering and organizing service for your home or office based in London UK. Our decluttering expert . . . offers expert advice and a hands on approach to dealing with clutter, whether you want to re-organize your wardrobe or blitz the whole house' (http:// www.organizerswebring.com/members/UnitedKingdom/england/england_ organizers.asp, accessed 24 November 2005). Part and parcel of this re-organization is the introduction of various little technologies, from storage devices such as baskets and boxes to 'expand-a-shelf organizers' that sit on shelves and, because they are stepped ('like stadium seating'), accommodate items (like tins or jars) which, as a consequence, become more readily visible and accessible.

Cwerner and Metcalfe note that decluttering rests on a model of domestic space that emphasizes channels and flows. The removal of clutter, and the ordering of things in their proper place, is thus concerned with keeping such channels open and enabling efficient flows of persons and things. In contrast, they show how such clutter plays important roles in the making of domestic space that is dynamic, contingent and iterative. Clutter is, as they put it, 'a social relationship uniting and separating people in time and space, a sense-making activity based on sensuous knowledge, embodied practice, informal routines and the structures of the life-world' (2003, p. 237).

Furthermore, such divergent views of clutter and tidiness map onto moralizing visions of technoscientific practice. For instance, the positive depictions of scientific practice found in the laboratories of the *Crime Scene Investigation* family of TV series present remarkably uncluttered and well-ordered domains in which, as we saw in Chapter 5, nothing technoscientific ever seems to go wrong. On this score, the decluttering of professional office life has been associated with the dream of the paperless office (*Economist*, 19 December 2002, http://www.economist.com/business/displayStory.cfm?story_id=1489224, accessed 14 November 2005). Accordingly, the morass of paperwork that goes into a particular task can, it has been argued (see Sellen and Harper 2003), be consigned to information technology. However, the filing systems people use on their computers can be hugely messy and ineffectual. Further, even if people use highly rational filing systems on their computers, there is no guarantee that people will remember where documents have been filed (this applies no less to physical systems).[11] As the *Economist* article goes on to suggest, the mess of paper – books, articles, notes – that clutters a desk does not reflect laziness or disorganization, but 'a physical representation of what is going on in [office workers'] heads'. In other words, physical clutter does have its practical uses. As Latour and Woolgar (1979) observed, the offices and labs of scientists are full of apparent clutter, and, as they note, such professional clutter reflects the heterogeneity and contingency that distinguishes scientific practice.

In relation to the present focus upon 'what goes with what' in the production of everyday space, we can note that, to the extent that mundane technologies (e.g. ICT, various domestic gadgets) serve in the moralization over, and the control of, clutter, this is not at all a straightforward matter. Such tidiness can generate disorder by virtue of unravelling the multiple connections entailed within clutter – multiple connections that enable people to deal practically with the tasks at hand. Put more strongly, we might say that practices, people and clutter 'go together' as more or less 'functional' spatialities that are disrupted by decluttering. Moreover, relating clutter to surrounding order and tidiness, we might say that 'looking down' into clutter, we find particular orderings, whereas looking down into the surrounding orderings, we find disorder, clutter and, indeed, chaos.[12] In the figure of clutter and in the promise of tidiness, we see two sets of relations – disordered and ordered – which on closer inspection turn out to be reversible: clutter becomes ordering, tidiness becomes disordering. In other words, we have complex interactions between divergent spatializations – or relations between relations – in which new sorts of spatializations are forged. This notion of 'relations between relations' is further illustrated and elaborated in the next section.

Driving through the borderlands

Michael (2004b) examines a largely unnoticed but everyday feature of driving, namely the animal roadkill that litters (one might almost say 'clutters') the roads of many western countries. Vehicles are responsible for the deaths of huge numbers of animals (see http://www.abdn.ac.uk/~nhi775/road_deaths.htm, accessed 17 October 2005), both in the countryside and in the city. In one sense, this is merely a continuing reflection of the centrality of motor vehicles in transportation – what has been called (the system of) automobility (e.g. Sheller and Urry 2000; Featherstone 2004; Urry 2004). In relation to everyday life, automobility manifests itself in a wide variety of ways that range from shifts in the experience of urban and motorway space (O'Connell 1998, Merriman 2004; Thrift 2004), through subcultural identification via particular forms of car consumption (e.g. Rosengren 1994; Lamvik 1996), to the 'immersions' between, and hybridities of, car and driver (e.g. Michael 1998b; Lupton 1999; Dant 2004).

However, such everyday movements, even within cities, are embroiled with animals and their mobilities – what Michael calls animobilities. This relates to 'animal geographies' that address the distribution of animals in relation to human societies. Animals in the guise of pets, laboratory animals, wild or feral animals and farm animals are thus situated culturally and physically in highly specific ways. According to Philo and Wilbert (2000), our views as to which animals belong in which sort of space (and with which sort of sociotechnical assemblages) are fairly standardized. They draw out the following cultural correlations:

> zones of human settlement ('the city') are envisaged as the province of pets or 'companion animals' (such as cats and dogs), zones of agricultural activity ('the countryside') are envisaged as the province of livestock animals (such as sheep and cows), and zones of unoccupied lands beyond the margins of settlement and agriculture ('the wilderness') are envisaged as the province of wild animals (such as wolves and lions). (2000, p. 11)

Now, as various authors note, these equations are anything but fixed. Obviously enough, there are 'impure' or hybrid spaces such as city zoos in which 'wild and exotic' animals are housed and subjected to what Franklin (1999) has called the zoological gaze. Moreover, such 'urban-wild' spaces are not uncommonly regarded as repositories of endangered species; in this respect, they are storehouses of a future wilderness. In other words, the urban comes to be seen as the potential saviour of the wild. However, as Sarah Whatmore (2002) has elegantly shown, the terminology of wild and wilderness must itself be problematized in that wild animals 'have been, and continue to be, routinely caught up within the multiple networks of human social life' (p. 9). If all this points to a picture of, to borrow from Whatmore, hybrid geographies, it is a picture full of ongoing tensions, not least because animals can travel across, and settle in, spaces where they once did not 'belong'. Out of such movements emerge what Wolch and Emel (1998) have called 'border-lands' in which 'humans and animals share space, however uneasily' (p. xvi) For example, as cities grow there have been increasingly frequent encounters between people and large predators such as cougars (see Gullo, Lassiter and Wolch 1998). In what follows, I briefly explore the borderland where auto-mobilities intersect with animobilities – that is where one set of relations relates to another set of relations – to create the everyday phenomenon of roadkill.

Now 'roadkill' is not in itself a transparent category. As Michael (2004b) notes, only a particular range of animals are common sensibly classifiable as roadkill (for example, such species as chimpanzee, tiger, gorilla, lion – what are occasionally called charismatic megafauna – are, in the West, more likely to be counted as the tragic victims of road accidents, than roadkill). Moreover, there are local differences in the way that roadkill is used, materially and semiotically: in some parts of the USA, roadkill is seen as potential food or as a means to education about local fauna; in some parts of the UK, roadkilled animals are casualties to be mourned, documented and rendered the subjects of political campaigns. These classifications of roadkill reflect differing spa-tializations of animobility and automobility. In the case of roadkill-as-food, the intersection of these trajectories is seen as 'natural'; in this figure coexist complex and ironic echoes of a culture that values hunting, survival and self-sufficiency. With regard to roadkill-as-casualty, this intersection is 'un-natural'; here roadkill signals victimhood born out of the encroachments of cars and their drivers into animals habitats (but also, out of much broader processes of encroachment that lead to habitat loss, which in turn makes it necessary for animals to use roads and road verges as wildlife corridors along which to move between habitat 'islands'). The hybrid character of the spa-tializations that emerge out of the relations between animobilities and auto-mobilities is further elaborated when we go on to consider some of the responses to this 'tragic' state of affairs. The use of various technoscientific artefacts – from signs warning about animals crossings, through tunnels or culverts via which animals might, it is hoped, avoid dangerous road surfaces, to modifications of the car to emit a high-pitched sound that frightens off hedgehogs – deepen the intersection between animobility and automobility (as does the fact that such mitigation is informed by data gathered by lay-people in the course of their daily journeys.[13])

It is clear that the relation between animobility and automobility can be described in terms of the idea of intersection or perpendicularity or, more simply, the preposition 'across'. And yet, animals make use of the road and its verges as was noted above. So, coexisting with perpendicularity are parallel relations between these mobilities (indeed, the mitigation measures we have mentioned are designed to reassert such parallel relations of non-interaction between animals and cars). Yet the motif 'parallel' does not really capture the fact that such non-interaction emerges in a spatialization in which animobilities and automobilities chronically intermix without perpendicularity. Road verges might have been created with cars and for cars, but, as with other artificial linear landscape features like hedgerows and ditches, they affect, and indeed facilitate, the movement of animals. Here, we seem to be encountering relationalities characterized by mobility and intermixing, at once parallel, a-mutual and separated, and perpendicularly and tangentially interlinked. Perhaps we need ways of capturing these relations between relations – that is, ways of addressing how, as with clutter, spatialities emerge out of the interactions of differing spatializations.

Conclusion: some thoughts on spatialities and prepositions

In this chapter, I have examined a variety of ways in which spatiality relates to everyday life and technoscience. Throughout, the complexity of space has been emphasized and we have examined how, within a range of everyday locals, we can detect the multiple signs of the global, which is itself seen to be local-izable. In the process, we have noted the contrary ways that technoscience enters into such spatializations of the everyday: mobile phones and telecommunications networks, maps and motor vehicles, domestic appliances and sunglasses – all these serve in the complex making of everyday spaces.

In the preceding section, on 'what goes with what', we have suggested that space might be seen as emerging out of the 'relation between relations'. Thus, in the case of clutter, we noted how the imposition of order (in order to make clutter fit in with surrounding orderings) could generate disorder, not least because clutter already contains its own peculiar order. In the case of roadkill, we suggested that the interactions between animobilities and automobilities were both perpendicular and parallel simultaneously.

Here, we might turn to Michel Serres' (Serres and Latour 1995; Serres 2003) call for a philosophy of prepositions. Key to Serres' interests is how disparate relations 'go together' – science and art, subject and object, the material and the semiotic – and how such going together makes order and disorder. In the process he has developed a variegated vocabulary through which he captures these heterogeneous relations and exchanges (e.g. Hermes, parasite, angels).

In relation to roadkill, Michael (2004b) suggests that the term frottage – the artistic practice perfected by the surrealist Max Ernst – might serve prepositionally to capture the simultaneously parallel and perpendicular relationalities of animobility and automobility. In frottage, matter (particles of the thing that is rubbed or does the rubbing, e.g. wood or paper or crayon) and signs (patterns on the paper, marks of the brass plates, blunting of the crayon) are transferred through the process of rubbing. The frottage of roadkill entails the rubbing of two 'surfaces' (that is, the relations entailed in animobility and automobility) and the transfer of matter (animal bodies, appropriate plants on verges) and signs (notions of species and corridor, signals of danger and care). Roadkill can thus be considered a moment of the frottaging of animobility and automobility. In a similar way, everyday clutter and tidiness entail the frottaging of divergent ordering and disordering relations that, through the process of frottage, at once transform and mimic one another. Of course, frottage is just one prepositional metaphor, and no great claims are made for it. Indeed, were we to apply the same line of thought to the disastrous interview episode described in the previous chapter, a different prepositional metaphor suggests itself. As we saw there, complex relations were enacted by various figures such as animals, technologies and human, combinations of these (pitpercat or intercorder), and their connections to and mediations of the 'global' (corporation and university sector). The succession and interdigitation of everyday spatiality entailed in the various figures that comprised the episode might perhaps be better captured by a notion like the maze: as we twist and turn within the episode, we encounter new relations. But with each such encounter, the maze reconfigures with new paths (and dead ends) made available.

Both frottage and the reconfiguring maze do not, of course, simply address spatial relations. They also incorporate temporal dimensions. It is to the temporal that we turn in the next chapter, where we examine how temporalities emerge in the complexities of everyday life and the dynamics of technoscience, in both of which we find, at once, routine and stability, and constant and dramatic change.

Notes

1. Of course, this self-sufficiency is hardly actual. Rather, it comprises an enmeshment in the complex circulations of materials and information that equip people with the wherewithal to do 'self-sufficiency' (see Michael 2000a; Macnaghten and Urry 2001).

2. An early example of a reaction against illumination is cited by Hughes (2004). In 1766, there was a riot against the Minister for War, Esquilache, in Madrid. As Hughes puts it, 'The oil-burning streetlamps that Esquilache had installed only a year before in the streets of Madrid, a brand-new civil amenity of which he and the king were quite reasonably proud – some 4,400 of them, each twelve feet high, iron and glass, all destroyed in a frenzy of direct protest, it seems, against illumination. To have light forced into dark corners was a sign that the king's ministers and his foreign bureaucrats did not trust them

[the people of Madrid]. It was an insult, like training surveillance lamps, automatic cameras, and microphones, after dark on today's city streets' (Hughes 2004, p. 50). This could readily be linked to the emergence of the panopticon as the structuring motif of modernity, and we could see a riot like this as a more dramatic example of the ruses and tricks that De Certeau detects in everyday life. However, we have touched on this model of space already, and in contrasting it to the Latorian notion of oligopticon and my own tentative notion of the taleidoscopticon, have pointed to some of its inadequacies.

3. This endless disclosability applies no less to the interior spaces of the psyche. Indeed, there are echoes here of the endless disclosability entailed in the transparency demanded by audit and accountability, as discussed in Chapter 4.

4. Informing this view is Whitehead's (1929) metaphysics of emergence: for 'space' we could substitute the heterogeneous prehensions that concresce to make an actual object or event – say, the particular use of a mobile phone. However, the 'nature' of those prehensions – of a particular space – means that not any combination or concrescence is possible. There is an element of teleology in play which means that for any actual event only certain prehensions can 'go' (or concresce or combine) with others. This is partly explored in the section 'What goes with what?'

5. For an extended example of the way that the 'environment' is made by walking through it, see Michael (2000b); see also Edensor (2000).

6. Carter and Michael's (2004) analysis is not directly concerned with the making of space: their emphasis on the performativity of the gaze serves to explore the epistemological role of vision.

7. The quote marks are added to 'visualization' to signal the fact that such representations are not simply visual – they entail a range of technologies of detection that, over and above ranging across the electromagnetic spectrum, also include the chemical, the physical (e.g. size, mass) and the aural. However, the final representation of such 'detection' will typically include a major visual component.

8. Running in parallel with environmental threat, and in some ways more evocative than it, is pandemic threat. Examples include HIV/AIDS, SARS and, most recently, avian flu. While such conditions are global, or represented as becoming global, they are often seen to originate in 'risky elsewheres'. Such elsewheres are accidental origins not only of disease entities, but also of risky biomedical practices that might generate combinations of medical and moral risk (e.g. xenotransplantation, organ stealing and trafficking, human cloning). If such concerns are vague and sensationalist, or mediated through rumour in everyday life, nevertheless, they can be said to map onto a spatialization where risks to morality (e.g. if it happens there, it will happen here) as well as health (e.g. the spread of mutated porcine endogenous

retroviruses) originate in elsewheres that are less subject to surveillance and regulation than we might expect 'at home' (see Dingwall 2001; Scheper-Hughes 2004). At the same time, such elsewheres have always been at 'home' too. We can note that technoscience, as a sometimes risky elsewhere, is itself spatialized in everyday life. There are particular spaces in which tech-noscience is seen properly to reside. One can conceive this as a nested – or, better still, patchworked – series of spatializations that map onto broader geographical categories (with all the political baggage that attaches to these): technoscientific innovation is thus located in the 'developed world' (e.g. the USA, Europe, Japan), in certain 'regions of the developed world' (e.g. Silicon Valley), in certain 'areas' (science parks, universities). In such local else-wheres, technoscience is often seen to proceed unchecked – motivated by scientists' unbridled ambition or curiosity (e.g. Michael 1992).

9. Once more we need to be careful not to romanticize 'community' as a co-herent entity (e.g. Cohen 1985). Inevitably there were those in the 'com-munity' who welcomed the airport development, not least as a source of economic benefit. Accordingly, it is important to consider the performative use of 'community' within the contest of the public inquiry in which claims and counterclaims are made about who speaks accurately for whom. Thus, spokespersons' claims for the community are as much about 'making' com-munity as representing it (in both senses).

10. Conversely, technoscientific readings of space can be shaped by everyday apprehensions. Anecdotally, an officer for an environmental organization once told me that his (and his colleagues') initial aesthetic reactions to a tract of land often motivated subsequent efforts to quantify technically its en-vironmental value. This is not a matter of contradiction or hypocrisy. Rather, it points to the complex interrelations between everyday and technical spa-tializations – interrelations that have been found at the heart of laboratory science (see Keller 1983).

11. We might also point out that ICT creates its own cumulative clutter – cables (for which Velcro® has developed tidies), diskettes and USB flashes, boxes and polystyrene. Thanks to Paul Stronge for alerting me to this additional mess.

12. Here, we can at once challenge and underline the chaotic model of the (so-cial) world in which islands of order float in a sea of chaos (e.g. Prigogine and Stengers 1984; Serres 1995b; Urry 2003). If clutter suggests chaotic islands floating about in, and hindering the efficiencies of, a surrounding order, looking down into clutter excavates how it serves in everyday ordering and how the surrounding order is in fact disordering.

13. The Mammal Research Unit of the University of Bristol in the UK (see http://www.abdn.ac.uk/mammal/roadkills.htm, accessed 9 March 2002) ran a Na-tional Survey of Wildlife Road Casualties and asked for volunteers to 'record wildlife road casualties observed during everyday journeys', including data about 'species, road cross-section, adjacent wildlife corridors (e.g. treelines,

water courses, headlands, etc.), blind bends, road verge habitat, adjacent land use, highway boundary feature (e.g. hedge, fence, ditch) and road category'. The aim of the research, which ran from June 2000 to May 2001, was 'to identify factors which cause wildlife to fall victim to vehicles in high numbers along certain sections of road ... [in order to enable the] ... design of road verge management prescriptions aimed at reducing wildlife road casualties'.

7 Technoscience, dis/ordering and temporality in everyday life

Introduction

At the time of writing (July 2005) Lance Armstrong has just won the Tour de France for the seventh successive time. He has, by many accounts, taken on 'mythic' status, not least because he overcame cancer, diagnosed in 1996, to become what is widely considered to be one of the greatest competitive road cyclists of all time. As an everyday public figure, he is renowned as much for his heroic determination – witness his cameo in the film *Dodgeball* — as for his extraordinary bodily capacities (it has been claimed that his heart is 30 per cent larger than, and his lungs twice as efficient at absorbing oxygen as, the norm). In his chosen sport and, in particular, his chosen event – the Tour de France – we find the juxtaposition of temporalities that has structured many a philosophical debate about time. On the one hand there is external or clock time (used as the objective measure of comparative speed) and, on the other, internal or proper (Nowotny 1994) or personal time (the experience of time in process or duration). Thus, when we watch the Tour de France – reputedly the most gruelling competition in any sport – we are witness to the contrast between the highly controlled movement, and exactitude, of the digital clock and the efforts of the riders focused on the process of cycling, a process in which time is experienced in complex ways (most obviously, its passing at varying 'speeds'). In the case of Armstrong, there is the additional layer to this experience: his cancer has 'framed' his life such that the time left to him seemed to be truncated. As many survivors have noted, one outcome of the process of survival is an effort to 'live life to the full' – to fill one's remaining time with as much value as possible, however that value is characterized (for related examples specifically concerning ageing, see Gubrium and Holstein 2000). Alongside all this, as we witness Armstrong's exploits on television, we experience the simultaney that is typical of many communications media (being in two places – the Tour and home – at the same time; see Kern 1983).

Armstrong can be regarded as, at once, an unusual physiological entity and an unprecedented historical figure. As an example of *Homo sapiens*, he can be seen to be, in some respects, a physically superior specimen within a population (for example, in relation to lung capacity). In our particular technoscientific era, this places him as a moment in evolutionary time (as opposed to, say, a god-given hero), whose uncommon physical attributes point to an uncommon genome and a brute advantage in the context of natural selection. As a 'sports personality', he connotes what we might call the 'end of history' for the Tour de France. In contrast to the evolutionary discourse in which we might well expect a super-Lance to appear at some point, much of the celebratory discourse around Armstrong points to his uniqueness: his record is unlikely ever to be broken; we shall not see his like again. Here, then, we encounter two different temporal narratives – both linear, but inhabited by different entities (biological bodies and genomes; sports personalities and heroes). Both are made eventful by the 'unexpected'; the former mediating the chance of genetics, the latter entertaining the vicissitudes of history. Here, we have touched on two different temporalities but, as we shall see, there are very many others (see Adam, 1998; Macnaghten and Urry 1998).

That Armstrong survived his cancer is down to, at least in part, the innovations of biomedical science. His 'untruncated' lifespan can thus be tied to technoscientific developments, developments that are continuing, as new diagnostics and therapies are pursued. In such survival we find interwoven another set of temporal motifs. While technoscientific developments have not untypically been linked to the idea of progress (and a linear movement of time forwards, towards more organization), cancer and prospective death points to increasing disorganization of the individual body, while death itself is traditionally thought of in terms of cycles of life and death. These three patternings of time – linear progressive, linear disruptive, cyclical – are also present in the Tour de France and its broadcast. The Tour itself has developed historically, not least in relation to the changes in the design and composition of the bicycles themselves (for an example of developments in cycles, see Rosen 2002). The Tour is structured by the diurnal cycle, though the watching of it is not (we can watch it at any time if we record it). As the race 'progresses', the order of racers becomes increasingly clearer, even as the bodies of the riders become increasingly exhausted (disordered). Within each stage of the race, the ranking of racers is constantly under challenge as individuals and teams jockey for position and attempt to rise up the order: sometimes this seems quite orderly, at others chaotic. Within all this is the routine practical relation between Armstrong's body and his bike. Here, for all the tactics and strategy that characterize a race as complex as the Tour, Armstrong must be 'at one' with his machine: it must respond to his body, and he must be corporeally sensitive to its capacities. This interaction is imbued with a certain 'automaticity' about it. That is to say, the tacit embodied skill that enables him to ride his bike, and the construction and design of the bike that enables him to ride, work seamlessly together to produce the rider-bike – an ongoing routine hybrid entity.[1] Here, proper time might be said to be tied to a hybrid or, rather, given that this 'automaticity' also rests on many heterogeneous relations to technologies and colleagues, tied to a sociotechnical assemblage.

These complex patternings are overlaid by others. For example, in relation to the ongoing enactments of the Tour de France itself, cyclical patterns can be detected in relation to the purportedly common taking of drugs that might enhance performance (progressive) but that might also disrupt the race (as famously happened in 1998 when the riders staged a sit-down strike). This is additionally complicated by the role of accusations about drug-taking – accusations that Armstrong has both delivered and received. Further, the race is punctuated by moments of chaos, notably accidents of various sorts which are now such an expected part of the Tour that they become almost 'routine'; in other words, disorder is almost an ordered feature of the race. On the above account, in the figure of Lance-Armstrong-on-the-Tour-de-France, we come across multiple, interweaving temporalities in which multiple and inter-digitating orderings and disorderings are played out.

Stephen Kern (1983) draws on the distinction between a future towards which we are more or less passively drawn and one that we actively pursue. In the context of his cancer, Armstrong has energetically supported various biomedical and public health endeavours, setting up the Lance Armstrong Foundation (LAF) in 1997, whose 'mission is to inspire and empower people with cancer to live strong. [The Foundation serves its] mission through education, advocacy, public health and research programs' (http://www. livestrong.org/site/c.jvKZLbMRIsG/b.695471/k.D29D/About_Us.htm, accessed 28 July 2005). As such, Armstrong is a medium and icon of hope and expectation about the future. Not only does the LAF provide practical support for survivors, it also promotes hope (see LAF's Tour of Hope at http://www.livestrong.org/atf/cf/{FB6FFD43-0E4C-4414-8B37-0D001EFBDC49}/ LAF AnnualReport03.pdf, accessed 28 July 2005) – an attitude considered to be an important component within cancer treatment regimes; see Brown 1998). At the same time, expectation is set up about technical advances that will 'beat the big C'. So, while technoscience might well serve in prolonging lifespans (though this is not uncontroversial: on the one hand, technoscience is also seen as being a source of medical risks, on the other, more mundane changes – in sanitation, for example – are credited with such prolongation), at the same time it is also tied up with the expectation of, and hope in, further advance. The spokespersons of technoscience, increasingly draw attention to what they expect to produce in the future. In this way, time can be seen to open up into a future that is constantly being drawn into what Nowotny (1994) calls an 'extended present' characterized by 'the mounting pressure that solutions to impending, recognizable problems have to be found now' (p. 52). Part and parcel of addressing such problems is, as we shall discuss in further detail below, representing the future: such futures of promise (e.g. a new treatment or a profitable innovation) are performative in that they are instrumental in making the present and thus, it is hoped, the future. They are a means of organizing resources, persons, materials, texts by persuading relevant others of the viability of particular futures.

The presentation of such representations of the future – their announcement – is itself a matter of timing, of choosing the 'right time', or 'kairos' as it is called in rhetoric (see Brown 2000). The notion of kairos allows us to draw one final observation from Armstrong's achievements. While kairos denotes the *choosing* of the right time – responding appropriately to circumstances – that is, exigence – we can also use this notion to connote the *making* of the right time. Armstrong has prepared meticulously for the Tour de France, visiting France frequently in order to analyse the route and practise on the particularly important sections of the course. However, this preparation needs to come together with a series of other factors: his physical and mental condition need to mesh with the sociotechnological assemblage that is his 'team' – an assemblage comprising, among many entities, other riders, bicycles, support staff and following vehicles. Moreover, kairos features in the timing of the announcement of his retirement: at the moment of winning his seventh Tour de France. At such a moment of triumph, he retires to spend more time with his children, suggesting that he is stepping out of the 'media spotlight'. Here, in the context of the temporariness of celebrity and the fickleness of the media spotlight, he chooses the kairos to make his announcement. But he has left a marker for a return to the media – not least, as some have suggested, as a politician (for hints of his political ambitions, see http://en.wikipedia.org/wiki/Lance_Armstrong, accessed 28 July 2005). The irony, of course, is that the stepping out of the media spotlight is conducted in the media spotlight; he has used global communication networks not so much to escape their reach, but to reorder his relationship with them – that is, to begin the process of redirecting his own biographical narrative (and thus to restructure his celebrity), something he is in a position to do partly because of the kairos of his momentarily intensified celebrity and the announcement.

The lessons of 'Lance'

For all the gaps, this has been a tortuous trawl through the times of Lance Armstrong. It has been my hope that, at the least, the foregoing gives some hint of the complex layerings of times even within a single, partial biography. Let me take a moment to abstract, and elaborate a little on, some of these features of temporality as they relate to the constitutive circulations of technoscience and everyday life (and the rest of this chapter).

Key among the range of narratives about time are those that represent sociotechnical progress in which we move towards ever greater achievement, not least in so far as prospective technologies are linked to a new future in which everyday life will in some way or another be transformed (e.g. the biomedical triumph over cancer). Technoscience, here, is tied to expectations about what the future holds in store – or, rather, the spokespersons of technoscience articulate particular futures that, in their performative aspect, are about ordering various actors (funders, regulators, publics) in the present so as to constitute the future. Needless to say, for want of a better term, such 'temporalizing accounts' are deeply contestable. As we shall see in the next section, these disparate accounts have a number of features that serve in rendering them more or less persuasive. In particular, we will examine those aspects of their rhetorical structure that serve in making futures (but also pasts and presents) attractive to a greater or lesser extent. We can note at the outset that these features of 'temporalizing accounts' are not restricted to technoscientific arguments about the future, but are a part of everyday discourse.

Technoscience has also impacted upon such narratives by playing a central part in the changing scales, tempo, units, exactitude and rhythm of the time of everyday life. Technoscience's constructions of, and interventions in, nature have expanded our temporal horizons into the distant past and far-off future, and set new limits on the beginning of the universe and the earth, and their end. Technoscience has entered into domains where things happen at incredible speeds (the nanosecond level) and slowness (the development of evolutionary trends, geological structures, star and planetary systems). If these representations describe the 'outer boundaries' of time, they circulate within an everyday life characterized by 'intermediate' temporal spans – daily, annual and life course cycles (as Lefebvre argued). These cycles would have had everyday technology at their heart – technologies such as the plough and the mill were central to the tempo of everyday life (e.g. Ingold 1993). However, modern technoscientific artefacts have impacted on these in seemingly new ways. Thus technoscience has been instrumental in the global standardization of clock time. As we shall see, from the invention of the clock to modern global positioning systems, time has become increasingly locked into a technoscientific system of standards that has entrenched still further the role of clock time in the structuring of everyday life. As such, 'normal' temporality in the contemporary everyday is thought to be rather different from that of preceding eras.

As noted, technoscience is at once linked to the reproduction of a 'time of normality' and the routine, and is instrumental in the production of novelty and transformation. Put very crudely, 'mundane' technology serves in the reproduction of the same; 'exotic' technoscience serves in the production of the new. Of course, the mundane and the exotic boundary is highly porous but, as we have already seen, narratives abound that make use of this divide (e.g. see Chapter 3 on mundane and posthuman bodies). It will be suggested, especially in the context of the 'extended present', that it is more useful to think of the novelty/mundanity in terms of complex patterns of ordering and disordering.

The penultimate section of this chapter deals with the individual experience of time and, in particular, the contrast between clock time and personal (or 'proper') time. Following Nowotny (1994), by proper time I refer to the relatively recent development of a sense of one's own time: 'self-time from the perspective of the individual' (1994, p. 41).[2] Initially this arose through industrialization with its imposition of clock time through which time became 'privatized' and exchangeable for, and convertible into, money (prior to that, time was 'public' – marked by common religious festivals, for example). According to Nowotny, the role of technology has been pivotal in the spread of proper time. Thus, for instance, the rise of telecommunications means that people are able to immerse themselves in their own 'private' or 'proper' time (e.g. watching TV, working on a computer). But as Nowotny is careful to note, such technologies are part of a broader complex of changes, not least those implicated in the processes of 'individualization' (in which, for example, people use technologies in the crafting of their own biographies). As hinted at in the discussion of Lance Armstrong, such proper time is shaped by various factors in everyday life, not least the sense of one's mortality – a sense that is bound up with technoscientific prospects and their interlacing with personal hopes. Yet, as we shall see, 'hope' is itself a complex notion that signals, at the least, both the everyday possibility of resisting or overcoming disaster and the potential of something like a 'better world' in which everyday life is itself transformed.

Temporalizing accounts: performing technoscientific futures

In considering what might be called rhetorics of the future, or temporalizing accounts, I draw on Michael's (2000c) analysis of a number of parameters along which such futures are represented. It will be more than apparent that these parameters are neither exhaustive nor discrete.

Given the way that modern westerners spatialize time (that is, array past, present and future along a line), it is unsurprising that representations of the future will implicate a particular *distance* from the present, and such distance enables the future to be 'performed' in particular sorts of ways. Thus, to argue that environmental catastrophe is in the far distance is to suggest that there is no hurry to implement remedial measures; to insist that it is in the near future means that action must be taken immediately. Inevitably such claims about distance invite counterclaims, not least in relation to technoscientific policy making. For example, genetic modification has been regarded as an improvement on evolutionary processes or traditional breeding programmes because, say, desired characteristics in crops or livestock are brought closer to the present. This future is worthy because it can obviate other futures in which malnutrition and starvation remain chronic. For critics, such initiatives generate future risks – risks that are more near-term than they need be – by, for example, bypassing long-term selection processes that afford the safety of such 'new species' (or new genomes). Further, the ways in which such 'distance' is measured – that is, units of time – can be used to rhetorical effect. To measure the arrival of the future in terms of weeks, elections or average lifetimes has different implications for the urgency of relevant action, the practicability of intervention or the likelihood of garnering political support.

Closely related to the distance of the future, is the speed with which we approach it. As Nowotny (1994) and others (e.g. Rappert 1999) have noted, we seem to be approaching the future at ever increasing speed.[3] In this context, depending on the sort of future envisioned, a speedy approach can be viewed as either deeply problematic or absolutely crucial. Conversely, slowness can be either the source of exasperation or a mark of due caution. It is easy enough to recount the preceding paragraph's example of biotechnology in terms of speed. However, in addition to accounts of the speed with which we move to *specific* technoscientific futures, there are also accounts that address the speed with which we travel towards the future per se. Increasingly, there seems to be a countermovement, as it were, to the speediness of modern everyday life in general. The headlong rush into the future – a rush partly enabled and mediated by technoscience – needs to be brought under control. Against the inevitability of speed, therefore, there is an alternative rhetoric of considered slowness, as evidenced in such books as *In Praise of Slow* (Honore 2005) or *Ten Thoughts about Time* (Jonsson 1999). In such accounts, speed is aligned with control – agency is regained as we practise various techniques through which we slow movement into the future. Re-reading this through Kern's (1983) distinction between grasping the future and letting the future grasp us, we can note that low speed or deceleration might be viewed either negatively as a form of inertia or reflex trepidation (e.g. not-in-my-back-yard), or positively as a manifestation of reflexive modernity (e.g. Beck, Giddens and Lash 1994). Speeding towards the future can likewise be interpreted in terms of the exercise of agency: negatively as a form of irresponsibility, or positively as an instance of entrepreneurial initiative.

Another dimension of temporalizing accounts concerns 'who' or 'what' is doing the movement towards the future. Such whos or whats might range from individual sufferers urgently in need of clinical options, through national economies that are obliged in the globalized marketplace to invest in particular technoscientific prospects such as nanotechnology or stem cells, to heterogeneous collectivities such as 'the earth' that incorporate ecosystems and future generations faced with global catastrophe. Here, such rhetorics can take on more or less standard narrative forms. Our three examples could, respectively, be narrated as biography, foresight analysis or doomsday scenario. Of course, in the cut and thrust of argumentation, such narrative forms are themselves open to debate not least because they invite identification (or afford subject positions for readers and listeners). Thus, to be moved by biographical narratives of the clinical needs of sufferers is to be less concerned with the possible negative outcomes for collectives: xenotransplantation might save the lives of identifiable individuals but at the risk of retroviral pandemics. This rhetorical terrain abounds with accusations of gross sentimentality and chilling cold-heartedness.

In arguments over the future and our movement towards it, assumptions are also made about the 'reason' for pursuing such futures. This can be conceived in terms of the extent to which the content of the future is interrogated. Often, it is claimed that what counts as 'good' or 'desirable' about a future is simply a reflection of what currently counts as 'good' or 'desirable'. For instance, posthuman futures can be said simply to be contemporary masculinist fantasies writ large (Hayles 1999); the futures of foresight analyses assume the continuing dominance of existing 'global market forces'. In these cases, the focus is upon the 'getting there' – a means-orientated instrumentalism whose focus is upon realizing unproblematized, that is, 'given', futures as efficiently or effectively as possible. By contrast, there are accounts that query the nature and substance of the future – for example, the meaning of the 'good life' (even though this will inevitably be constrained by current thinking on such matters). Much of the technoscientific argumentation about the substance of futures gravitates around issues of realism and realizability. For example, certain accounts of environmentally friendly futures might be subject to accusations of utopianism. And yet, environmental activists might counter that it is their 'utopian' futures that are really 'realistic' if humanity is to survive sustainably. All this is intimately tied up with issues of 'hope' and I will return to this at the end of the chapter.

Finally, there is the most obvious aspect of the future – its valency. Whether it is a bad or good future is obviously a partial upshot of all the other parameters listed above. And, of course, there is a considerable history of political discourse, not least on utopias and dystopias, which informs the western sense of a good or bad future. However, there are also other *formal* aspects of the way that the future is represented that affect its perceived valency. For example, the certainty – or modality (Latour and Woolgar 1979) – with which a prospective future is articulated can be seen as reassuring or suspicious. To insist that a particular problem will *definitely* be solvable in the future might be greeted with relief or incredulity. Having noted this, in the present climate of public scepticism in the UK, scientific institutions whose pronouncements on some state of affairs are couched in the terms of certainty, are liable to be met with intense scepticism as a matter of course. Moreover, such reactions are themselves open to similar interpretation – public scepticism entails a certainty that is itself suspect (see Michael and

Brown 2005). More generally, if positive accounts of the future can be criticized as 'talking up' the future for the purposes of generating optimism, or hope, or support, these critical accounts can themselves be criticized as indicative of conservatism, or pessimism.

This last set of concerns points once more to the complexity of the performativity of technoscientific futures. First, while we have focused here on the content and style of the arguments, these must be put into circulation – e.g. through the arrangement of press conferences or embargoing of press releases. The timing of such 'putting into circulation' – the kairos – is, as suggested above, part of the rhetorical potency of such arguments. As Brown (2000) has shown in the case of Dolly the cloned sheep, if the appropriate conditions are not in place, the effect of one's announcement can be badly misjudged, resulting in unexpected levels of public reaction and media scrutiny. Second, such performativity is also linked to a range of heterogeneous circumstances – an array of other discourses, practices, materials and relations. In the case of xenotransplantation these might include unauthorized filming in research institutions, the relocation of researchers to more conducive national regulatory environments, the organization of regulatory public meetings, the recalcitrance of porcine and human bodies, the caution of venture capitalists, and the media by which these events are rendered and circulated, not least mundane technologies such as telephones and cars. Kairos here is a complex and possibly fortuitous accomplishment. Third, as should be obvious, such rhetorics of the future entail particular readings of the past and the present against which the future can be contrasted as a positive advance, or, alternatively, a steep descent. Needless to say, just as the future is contestable, so are these readings of the past and the present. The persuasiveness of particular futures is thus hugely dependent upon the persuasiveness of their related pasts and presents (and, of course, of the linkages between these – that is, of the narrative through which past, present and future are connected).

I have suggested above that the performativity of futures can be thought of in terms of the attempted ordering and disordering of a variety of different actors and entities: aligning or disaligning them so that one's hoped-for present is established, rendering one's hoped-for future more likely. In the case of xenotransplantation, advocates (e.g. PPL Therapeutics plc, press release, 11 April 2001) have represented the future of organ transplantation in terms of an inexhaustible and global supply of organs (where now there is a worldwide shortage) that is at once ethical and technically unproblematic (hence the choice of the pig as the 'donor animal'; see Brown and Michael 2001), and profitable to boot (the potential market was said at the time to be worth $10 billion). In the context of such a future, contemporary efforts at developing xenotransplantation appear sensible in the extreme, and fully deserving of every support. However, members of the public have a very different view of the future. For patients, whose conditions might well benefit from xenotransplantation, such futures are seen as desirable but in a cautious way. For such actors, positive changes to their everyday life (and its prospects) are desirable, whatever their source. Rather than investing towards a singular future implied in advocates' arguments, patients want a plurality of futures – many initiatives need to be supported in the hope that at least some will produce tangible benefits. The anxiety is that scarce resources that could go into a variety of projects will be monopolized by a few.[4] In other words, there is a desire among patients for looser ordering – or for multiple orderings to be in process (Michael and Brown 2003). This is tied to another concern expressed by patients, namely that announcements of breakthroughs or new discoveries are often tied to unrealistic claims about imminent treatments. Patients have plenty of unhappy experience of biomedical announcements that pointed to a rosy future but that subsequently failed to deliver. For the patients it was more important that such announcements were not premature – that the future they were promised was very near indeed. One could almost say that, for these patients, there is hardly ever a kairos for such announcements. This was because, announcements of breakthroughs that were ordering for advocates, were thus, ironically, very likely to be disordering for patients. In the context of everyday life with particular chronic conditions (e.g. diabetes or coronary heart disease), there is a delicate balance to be struck between hope and resignation (and the various practices of living that mediate this balance), and such announcements can all too easily upset that balance.

This section has attempted to document some of the ways in which technoscientific futures (and, of course, these need not be technoscientific – they can be political, or economic, or religious) are rendered persuasive. However, we have also seen that such futures are chronically contestable not least because, in so far as they are a means to ordering the present (in order to order the future), they are also involved in disordering. Those who are disordered will not necessarily accept this without some resistance or complaint.

Technoscience and changing (or not) the everyday

In what follows, we examine the way that technoscience has simultaneously generated change and sameness. Here we begin to see how time as represented (e.g. various futures) folds into the temporal doings of everyday life, doings that entail technoscience in further complex patterns of disordering and ordering.

Our everyday practical experience of time has, arguably, been transformed over the course of the 'modern' era. If once such experience was structured by 'natural' rhythms of days and nights and the coming and going of the seasons, these have become reconfigured – needless to say, technoscience has played a major part in this. Of course, such rhythms still shape everyday life but their role has become far more contingent. Thus electric lighting extends activity beyond diurnal rhythms; as we have seen, the spread of air-conditioning systems mean that the traditional tempo of work and sleep no longer need to hold; transportation and storage systems mean that we have ready access to produce that is 'out of season'.[5] The rise of telecommunications technologies has been particularly important in the erosion of temporal difference: as Kern (1983) notes, we are now in the thrall of simultaneity (Nowotny 1994) – that is, being in two places at the same time (or experiencing two times in a single place). When we talk on the phone, when we watch TV coverage of a distant sporting or political event, when we pick up our emails, we are at once in our local time and the time of an elsewhere. Technology as a part of the complex process of industrialization (which has included, for example, dramatic increases in mobility and the convertibility of time into money, as well as the romantic reaction against such changes) has also impacted on our sense of self time, or proper time as Nowotny has called it. And, of course, we have seen how the future, and the speed with which we move towards it, has been reconfigured – what Nowotny has dubbed the 'extended present' and what is popularly called 'time squeeze'.

If all this suggests a sort of normal 'temporal agitation' in which many times, rhythms, linearities coexist within everyday life, this occurs within a context of 'mundanity' and 'routine' (and, indeed, 'security'). The backdrop to such 'temporal agitations' is the ostensibly stable, and predictable, character of everyday life. For instance, time itself is standardized and stabilized. Although this was an enormously complicated sociotechnical accomplishment, we are now secure in the assumption that the unit of time does not vary (a second is a second everywhere), that time across locations is predictable, and that the time locally can be determined with relative precision. The role of technoscience in all these has been multifarious. For instance, the second now is very precisely defined – as it states on the US National Institute of Standards and Technology's Physics Laboratory's (Time Division) website:

> The cesium atom's natural frequency was formally recognized as the new international unit of time in 1967: the second was defined as exactly 9,192,631,770 oscillations or cycles of the cesium atom's resonant frequency, replacing the old second that was defined in terms of the Earth's motions. The second quickly became the physical quantity most accurately measured by scientists. As of January, 2002, NIST's latest primary cesium standard was capable of keeping time to about 30 billionths of a second per year. (http://physics.nist.gov/GenInt/Time/atomic.html, accessed 2 February 2005).

As has long been noted for processes of standardization in general, standards are maintained by a complex technoscientific infrastructure traced out by the circulation of standards and arguments about standards (see O'Connell 1993; Barry 2001). Such standards – the second, the volt, the metre – permeate everyday life, normally underpinning a routinized bedrock of mundane practices. It is not difficult to find examples of the way this process of standardization enters into more and more aspects of everyday life. For instance, in relation to the standardization of time, clocks can now automatically synchronize themselves with standard sources: in the UK, Oregon Scientific's range of radio control clocks have their time set by radio signal derived from the atomic clock at Rugby.

The broader point is that everyday life is littered with bits of technoscience that make it, at a practical level, predictable and routine, even if what it is that is predictable and routine has changed. According to Hillman and Gibbs (1998) in their popularizing book *Century Makers*, there are many 'clever things we take for granted' that have 'changed our lives over the last one hundred years'. Included in their list of such clever things are: cat's eyes, the Post-It® Note, the paper clip, the zip/zipper, the ballpoint pen, the ring-pull/pull-top, velcro, the child-resistant cap. These clever things – the products of technoscience – are, by and large, integrated into the fabric of everyday life. They are mundane, near-invisible artefacts that mediate the routines of everyday life in the home, at work, at leisure. They are, to re-quote Michel Serres, the objects that make 'our history slow' (1995b, p. 87). This they partly do through their operations upon bodies as they go about their everyday business of opening tins and dipping teabags, switching lights on and off, zipping and unzipping, and so on and so forth. Of course, as pointed out in Chapter 3, such 'everyday bodily business' is the business of particular bodies: such technologies discriminate – or rather, serve in the discrimination – against particular bodies.

The contrast that has been developed here is between, on the one hand, technoscience that is embroiled in more or less dramatic change and, on the other, everyday technologies that 'simply' reproduce certain everyday routines. So, iconically, we find information and communication technologies and new genetic technologies that promise, or are apparently in the process of delivering, major transformations in everyday life (not least in how we think of, and act towards, the body). By comparison, mundane technologies mediate more of the same.

As we have seen, however, technoscience that from one perspective seems to be revolutionary is, from another, often regarded as reinforcing existing patterns in everyday life. Contrariwise, as constituents of socio-technical assemblages, everyday technologies do not 'merely' reproduce the mundane and the routine, but can enable the emergence of more or less novel practical functions and enactments of self. The door closer and the electric toothbrush are, along with, respectively, stopping draughts and enabling dental hygiene, are also opportunities for making and remaking identities and social relations. For instance, sometimes, by virtue of accidental juxtaposition (what Michael (2000a) calls co-incidentalization) of different sociotechnical arrangements, such technologies start doing unexpected things. The TV remote control keeps getting lost down the back of the sofa (we might call this misbehaviour) by virtue of the fact that the designs of both remote control and sofa incorporate structuring assumptions about the nature of the hand. This leads to such innovations as new technologies that can detect the lost remote (another example is the 'locate' button on the base cradle of some cordless home phones), or new social routines whereby the remote is always returned to the same site (e.g. the arm of father's armchair; see Morley 1992).

The basic point here is that mundane technology does not always mediate routine orderings, and that 'exotic' technology does not necessarily yield disorderings and reorderings. So, it is neither the case that mundane technology can be unproblematically associated with a cyclical temporality (more of the same), nor the case that 'exotic' technology necessarily features in only linear temporality marked by change (whether to the 'good' or 'bad', to the more or less ordered). Moreover, when mapped onto each other, exotic and mundane technoscience generate at once change and stability. Let us consider an example.

Novas and Rose (2000) provide an account of the 'somaticization of personhood' in their study of a web forum discussion group concerned with Huntington's disease. This they see as associated with broader changes such as 'recent developments in the life sciences, biomedicine and biotechnology are associated with a general "somaticization" of personhood in an array of practices and styles of thought, from techniques of bodily modification to the rise of corporealism in social and feminist theory and philosophy' (2000, p. 491). The rise of various genetic technologies such as Polymerase Chain Reaction, or PCR (see Rabinow 1996), has enabled the rapid development of genetic diagnostics. However, diagnostics have generally outstripped the development of therapeutics: to be diagnosed with a particular genetic 'condition' (which might mean exhibiting no symptoms for many years) does not mean that this condition is treatable. The result is that the decision to submit to genetic testing generates major ethical and social ramifications for oneself – ramifications that partly reflect an emergent complex politics of genetic risk (Lemke 2004). Such a somaticization takes place within, and impacts upon, the context of other relations that are both altered and reproduced – relations with experts, relatives, friends, fellow 'sufferers'. In addition to a web forum through which some of these relations are mediated, more modest mundane technologies are also involved – phones, cars, shoes, and so on. Here, then, new and old relations are being enacted – that is, there is a multiple temporality in process in which new somatic identities, and existing identities of parental responsibility, friendship and consumption are simultaneously performed. In the context of these emergent discourses and practices, mundane technologies themselves become 'strange', just as in the context of existing discourses and practices, new technologies (the web forum) are strangely familiar. The phone becomes not only a mode of commonplace communication, and thus a continuing part of a stable sociotechnical assemblage, but also a bit of technoscience that materially and culturally serves in the performance of 'somatic identity'. In sum, mundane technology is working upon, inflecting with and mediating exotic technoscience (and vice versa), and in the process the 'temporal performativities' of both are being altered. That is to say, the phone serves in the making of change and non-change, as do genetic diagnostics and the web forum. By the term 'temporal performativities', I mean to suggest that these technoscientific artefacts, as part of sociotechnical assemblages, mediate both change and non-change, and are themselves both stable and unstable. They at once retain a certain character but they are also 'in process' – emergent – and thus should not be imagined to be simply 'there' acting in the 'usual' ways.[6] It should be clear that such complex performativities with their multiple temporalities entertain both human and nonhuman components – they are hybrid. Here, then, as with Armstrong-and-his-bike, there is mundane routine, as with the dynamics of a Tour de France stage, there is an interplay of order and disorder and, as with the Tour de France as a sporting event, there is the making of history.

Temporal performativities and the everyday present

These temporal performativities, as with the rhetorics of the future (which are a subset of temporal performativities), are concerned with constituting the everyday present in order to effect a particular future. Here, I follow the philosopher Michel Serres (1995a, 1995b; Serres and Latour 1995) and thus want to avoid imagining time primarily as a moment of change or sameness – that is, a 'present' – that is a point on a line, where that line is time stretched out linearly over space. As such, time should not be regarded as transcendental, objective, external. I also do not want to think of time as a sort of rolling subjectivist moment in the mind that draws on the past and projects into the future (as might be associated with Mead or Bergson). Instead, along with Serres, I aim for a model of time in which the present moment is emergent in the processes of connection and flux. According to Lash, Quick and Roberts (1998), Serres regards time as 'a flattened and immanent scenario of connections, fluxes, and objective intensities' (p. 4) where 'the present is singular' and is seen as 'a given assemblage of particularities' (p. 4). These particularities include human and nonhuman, subject and object.

So, the present might be thought of as made up of connections, fluxes, movements, communications that are ordering or disordering in various ways – that is, entail particular patternings or trajectories. Thus, there are some patternings that are negentropic in which out of chaos or disorder emerges some new order. There are some trajectories characterized by a seeming stasis or lack of change in which there is a continuity of order – that is, a homeo-static process in which a particular pattern is ongoingly reproduced. Finally, there are those presents in which the movement is towards disorder (or chaos) – entropic trajectories typified by disruption of existing orderings. A key point is that all these can coexist; one of Serres' long-standing concerns is the re-lation between order and chaos (Latour 1987b), and the processes whereby ordering and disordering both occur simultaneously and emerge out of one another.

So, temporal performativities are part of the process of making such complex patternings – patternings that are given to contestation. As we saw in relation to the discussion of rhetorics of the future, temporal perfor-mativities that seem to order the present (e.g. for advocates of xeno-transplantation) are disordering for, and potentially resisted or subverted by, others (e.g. patient groups, who promote alternative patternings). We have further attempted to show how these orderings entail heterogeneous socio-technical ensembles (in which seemingly stable mundane artefacts like the telephone might be said to change). To put this another way, such ensembles are characterized by their own proper time, which, when viewed as perfor-mative (i.e. as temporal performativities), also effect – at least potentially – the proper times of other ensembles (which are, of course, engaged in their own temporal performativities). There is no abstract or universal time 'within' which these interactions take place – rather, temporality emerges in the interactions between these ensembles' performativities.

In order to illustrate this account of competing temporal performativities further, let us consider the everyday activity of queuing. Of course, this is not a uniform activity – there are many different sorts of queues. There are queues at bus stops, train stations and petrol stations, in shops and supermarkets, in cafes and restaurants, in doctor's waiting rooms and hospital accident and emergency departments, on the telephone and on the computer. In the context of 'physical' queuing, there are several features to note. First, as various commentators have recognized, queuing is an everyday activity that is deeply familiar to participants; even if its 'rules' cannot always be articu-lated, it entails constant awareness of such facets as the changing order of one's position within the queue – an awareness that is a 'moral requirement' (for example, Laurier, Whyte and Buckner 2001; Garfinkel and Livingston 2003; B. Brown 2004). Second, within these expectations of queuing, there are ways in which people can warrantably break the rules and jump the queue. In the case of ostensible emergency, they can push to the front, or be ushered to the front as is the case in A&E (see Carlin 2003). The same applies when, for instance, the end function of the queuing for the would-be queue jumper is less arduous – in the sense of being more 'automatic' and thus requiring less attention at the queue's end point. Thus, as Normark (no date) notes in re-lation to paying at petrol stations, customers who are paying by cash for petrol only (as opposed to paying by credit card or for petrol and groceries, can bypass the queue and leave the money on the counter). In the seemingly simple process of queuing, there are numerous temporal performativities and complex orderings to be unravelled.

Here, drawing on many hours of undesired and unintended participant observation, I explore the example of a busy train station at which there are several queues – a long, snaking queue framed by a taped barrier to a number of ticket kiosks, a short queue to two information and advance booking desks, and two other separate queues to automatic ticket machines. The obvious initial decision is which of the ticket queues to join – a choice that is shaped by a range of considerations, such as the length of the queue (something that varies during the year – the main queue is much longer during the school holidays), experience of the relative rate of movement of each queue (the ticket machines are not always reliable and require skills that vary across customers – so the machine queues, while shorter, move more slowly and/or erratically), expectations about the punctuality of the trains (can one risk the longer staffed queues on the basis of the likely lateness of a train?) Here, there is a complex choice of temporal performativities – the choice aims to con-stitute an ordered present that will effect a particular ordered future (catching the train on time). However, the customer is also entering into the temporal performativities of a sociotechnical assemblage – one that encompasses such elements as the timetables and their shortcomings, and the staffing policy (time equals money) of the rail company, as well as the queue itself.

Within the queue, as noted above, there are certain 'rules' that allow for the doing of queuing (or 'working the queue'), rules that also allow for their own warrantable transgression. The orderliness of this sort of queue as a collective temporal performativity is always in danger of being disordered – there are always minor infringements and repairs going on. Routine enact-ments can always be derailed and sometimes there is a failure to meet the 'rule' of promptly moving up the queue as places are vacated: queuers might be distracted from attending to the queue's movement by friends or relatives, or they might be hampered in their progress along the queue by their own cumbersome luggage. The latter little disorder is regularly repaired through the help offered by other queuers; or there might be some sanction levelled (say, in the form of various gestural and vocal signs of exasperation). These assistances and sanctions are also enacted where those at the front of the queue fail to respond to the signals – illuminated numbers above the desks – indicating that a ticket seller is free, or when the customer, on reaching the desk, takes too long a time (e.g. because they have failed properly to prepare themselves to conduct the transaction). Here, the little temporal performativities of individuals and groups (say, those who have arrived in good time at the queue) intersect problematically with those others (say, those who have not, and for whom the swift movement of the queue is of higher priority). Of course the obverse occurs too: those whose temporalities are less urgent indicate their disapproval of others' 'impatience'. Here, then, within the queue are vying temporal performativities in the process of 'calibration' or 'management' that make the queue 'work'. Additionally, there is an ongoing sensitivity to the various temporalities mediated by the sociotechnical assemblage of the railway station or train timetables. As an ongoing process, various aspects of the sociotechnical assemblage of the queue are taken into account. For instance, if a ticket desk closes, or a particularly popular train is announced as late, then this might engender a sort of further calibration in which performativities align around, respectively, greater or lesser urgency.

This has implications for other sorts of rule breaking. As we might expect, the 'no-jumping-the-queue' rule is also broken. Some customers will sometimes make their way directly to a ticket desk. This is warrantable if certain apologetic requests are properly made, indicating either urgency or, more often, less arduous end functions (e.g. seeking some quick advice or picking up a waiting ticket). However, the permissibility of this rule breaking also rests on the perceived 'urgency' of the queue: where it has been long or slow, or where there has been a closing of ticket desks, queuers-in-a-hurry might feel they are in a position to enforce the rule of no queue jumping, irrespective of queue jumpers' warrants. Even those not in a hurry might under such circumstances feel obliged to resist queue jumping for the sake of other queuers-in-a-hurry. We might say that the warrantability of refusal to allow otherwise 'justified' queue jumping requests is thus affected by the state of the queue; alternatively, we might say that such refusals are part of the temporal performativity of the queue as a sociotechnical assemblage that incorporates such elements as timetables, numbers of open ticket desks, the efficiency of the credit card system and the supporting networks, alternative ticketing arrangements, staff shortages and labour relations, and the chronic sociotechnical fallibility of the British railway system.

It is hoped that this last example illustrates[7] how different personal times, as temporal performativities, compete and collaborate in the emergence of the sociotechnical assemblage that is the queue-at-a-railway-station. In the process, the interactions of such temporal performativities serve in the enactment of the complex shifting temporality of the queue itself. Put another way, we have tried to trace some of the many ordering and disordering 'moments' that make up the more or less orderly – that is, homeostatic – queue; 'moments' that include not only the personal times of the queuers, but the temporalities of such ensembles as credit card processing systems, timetable display screens, and the condition of trains and rails.

Conclusion: hope, or the temporal performativities of new (or better?) orderings

The preceding examples have been used to describe one particular way in which we might conceptualize temporality in relation to technoscience and everyday life. The present analysis has emphasized that 'time' entails 'struggle' and, as such, relations of power are part and parcel of the various competing temporalities we have considered above. In other words, the idea of differing temporal performativities is fundamentally concerned with the way that actors make particular presents, which in waylaying or disrupting or curtailing others' presents, serve in the possible production of 'future presents'.

However, it might be argued that such everyday struggles over time are informed by, or contain within, the kernel of a 'better time'. Thus, in relation to the queue, a 'better time' might be one of temporal plenty – where there would be no hurry, where there were ticket booths and trains to accommodate everyone in good time (or, rather, public transport was free), and where any queue was more carnival (as Lefebvre might put it) than a more or less tacit race to the front. As in the dole queue in *The Full Monty*, where the protagonists together practise their dance moves, the queue can be turned from a strategy wherein bodies are (within certain limits, as we have seen) disciplined by the sociotechnical assemblage that includes clocks and timetables and tape barriers and computers, to a setting where tactics that resist such discipline are played out. Here, such tactics might well be 'nostalgic' – De Certeau, Giard and Matol's (1998) account of everyday tactics informed by 'local tradition' (see Chapter 2) can be regarded as, at least in part, entailing nostalgic evocation. But they can also be seen to contain an element of 'hope'. It is instructive here to draw briefly on Kellner's (no date) reading of the work of Ernst Bloch, the utopian Marxist philosopher. For Kellner's Bloch, the present can be understood in the context of the past, understood as a 'repository of possibilities' (p. 2), which can serve in the identification of possible futures. The 'present moment is thus constituted in part by latency and tendency: the unrealized potentialities that are latent in the present, and the signs and foreshadowings that indicate the tendency of direction and movement of the present into the future' (Kellner, no date, p. 2). To access these emancipatory potentialities, latencies and tendencies, there is need of a 'dreaming forward' in everyday life – what Kellner calls an 'anticipatory' dimension that entails utopian images of a better world. This version of everyday life as 'positive critique' we have already encountered in the work of Gardiner (see Chapter 2).

There are a number of points to make about this. First, there is no attempt in this sort of analysis of everyday life to detect a fully blown vision of a utopian future world. This is, as Garforth (2005) has noted, symptomatic of the way utopias are nowadays regarded. Even though within contemporary utopian literature future societies are still vividly articulated, they have been thoroughly problematized by analysts (not least in terms of their historical contingency). Indeed, their function might better be regarded less in terms of setting out a blueprint for an ideal society, as about 'transgression, estrangement and otherness' (Garforth 2005, p. 9). That is to say, while still 'linked to hope and future possibilities' (p. 9), the focus is upon the *process* of transformation, rather than the end point.[8] Now, and this is the second point, according to Cohen and Taylor's (1992) *Escape Attempts*, such processes of 'transgression, estrangement and otherness' in everyday life are deeply problematic. Forms of estrangement such as distancing oneself from what they call 'paramount reality' might well serve to reproduce that reality. Thus being distanced from one's role as a particular functionary in an institution (e.g. insisting that one is more than one's role) does not mean that one performs any less efficiently. However, in the Introduction to the second edition of their book, Cohen and Taylor reflect on the fact that there might be no such thing as a 'paramount reality' to be resisted (or escaped from). In such a case, what it is that is being transgressed and what might count as otherness become altogether more slippery.[9] Indeed, the 'reality' that has been described above is one of multiple temporal performativities – there are many 'paramounts' (and resistance to these 'paramounts') within it. Moreover, these are temporal performativities enacted by combinations or assemblages of humans and technoscience, in which both humans and technoscience are emergent. Third, this last point suggests that if we want to retain a notion such as 'hope', we will need to rework it as a temporal performativity that applies to emergent assemblages. In other words, we might have to think of 'hope' as something like a quality of the temporal performativities of socio-technical assemblages. The orderings/disorderings of such assemblages generate possibilities and differences – 'moments of presence', to borrow from Lefebvre – in which past, present and future performativities combine anew in a 'heterogeneous dreaming forward'.

Notes

1. Here, proper time is tied to a deep-seated practical 'familiarity' with technology – indeed, proper time might be seen as a property of the hybrid of person-and-technology. For an example of a hybrid analysis related to the bike, see Rosen (1993). For an example in relation to the car, see Michael (1998b).

2. Proper time needs to be differentiated from the idea of the processual aspects of experiential time that we find in Bergson or Mead. For Bergson, such

experiential time – what he calls duration – cannot be truly represented. One (inevitably spatial) metaphor that has been used refers to a continuous flux in which there is a succession of states, and where preceding states merge into subsequent states. However, rather than see these states arrayed in space, they should be imagined from the singular position of the movement. (For an important commentary on this, see Ansell Pearson and Mullarkey 2002). Whereas duration is a universal property of consciousness, proper time, by comparison, refers to a historically contingent form of experiencing time.

3. Michael (2000b) argues that the potency of this rhetoric of speed is partly rooted in the modern western 'love' of speed (see Millar and Schwarz 1998) – a love partly rooted in various technologies of movement (including tele-communications). Of course, even in the modern West this 'love' is hardly universal.

4. This is but a subset of the temporalizing accounts that one might find among non-experts. Michael and Brown (2003) found that, among members of the public without any particular interest in xenotransplantation, there was an account of the movement towards the future that was in others' – often, experts' – control. That is to say, there was no expectation that one would directly influence the course of events. Nevertheless, within the immediate context of talk (a focus group), such accounts served in the performance of self as 'realistic'. Another pertinent temporalizing account is the 'slippery slope' discourse – a change in a particular direction that will precipitate an avalanche of more dramatic changes (see also Mulkay 1997). An instance of this might be found in the recent consultation exercise by the UK Government's Department of Health to review the Human Fertilisation and Embryology (HFE) Act 1990 – the law regulating embryo research, fertility treatment, gamete donation, etc., in the UK. One of the issues under consideration is whether there should be sex selection in IVF treatment in order to 'balance' the sexes of children within families. This was a change recommended in the House of Commons Science and Technology Committee's (24 March 2005) report on 'Human Reproductive Technologies and the Law', and to which the government, in its Response (August 2005) was negative, though said it would 'seek wider public views' (p. 19). As noted in *The Times*: 'Critics fear that sons may be favoured over daughters and predict that the move [to sex selection] will make it harder to prevent the selection of traits such as looks or intelligence, should the technology become available' (http://www.timesonline.co.uk/article/0,,2-1738511,00.html, accessed 17 August 2005).

5. As noted, such contingency is itself open to critique, not least around issues of the environmental sustainability of sociotechnical practices that overturn, or at least sidestep, 'natural' rhythms.

6. In their resilience and continuity, one might say that such mundane artefacts are 'stubborn' (to borrow from Whitehead), but in relation to their emergence

– to the fact they can be seen as changing as they incorporate other functions and uses (or prehensions as Whitehead might say) they, like the laws of nature, should not be subjected to the fallacy of misplaced concreteness (see Whitehead 1925).

7. I say 'illustrates' because certainly my account is not meant to be exhaustive. For instance, the temporal performativities that comprise the queue also abound with prejudices around national stereotypes (e.g. certain nationals are sometimes thought to ignore the protocols of queuing, while others are thought to be obsessive about queuing). There are also comparable clichés that are sometimes enacted in relation to older people.

8. Of course, the genre of science fiction has contributed to this utopian literary tradition (as well as providing numerous dystopias). It has been a key resource in the advocacy and critique of various futures – complex motifs such as those of Frankenstein have affected the sorts of future people fight for or against (e.g. Turney 1998). In the present context, I note only that everyday life can be thought of as 'hopeful', and that this hope might well be informed by utopian writings, and specifically science fiction (and also more journalistic projections of the technoscientific future such as that with which I began in Chapter 1).

9. This is especially problematic taken in the context of consumer culture, where 'otherness' can be promised by consumer goods – that is, can itself become commodified or domesticated.

8 Conclusion: questions of technoscience, everyday life and identity

The woolliness of the everyday

Post-it® Notes and Teslar watches, mattresses and antisnoring devices, sunglasses and electric toothbrushes, vacuum cleaners and traffic surveillance cameras, tape recorders and mobile phones, maps and home diagnostic kits – these are some of the 'singular' everyday bits of technoscience that populate these pages. As we have seen, people enact themselves – their bodies, their politics, their socialities, their spatialities and their temporalities – with the aid of these disparate artefacts. Or, rather, 'people' emerge in their corporeal, political, social, spatial and temporal complexity and multiplicity in relation to these artefacts. Of course, these technologies are hardly singular or stand-alone, but are, as we have insisted throughout, themselves emergent from sociotechnical assemblages. We have explicitly considered a number of such assemblages: 'railway station queues', the Tour de France, 'traffic surveillance', 'xenotransplantation', 'the research interview', *Crime Scene Investigation*, 'clutter' and 'roadkill'. These assemblages have entailed patternings of ordering and disordering marked by various contestations and collaborations that entail temporal performativity and the enactment of potential futures.[1] We have also touched on more 'systemic' technoscientific configurations that have 'gathered up everyday life', as it were, and are reputed to characterize the present (or almost present) epoch. Virtual society, surveillance society, network society, geneticized society, posthuman society, biosociety – all these have been said, in one way or another, to capture the key dimensions of the current era. In contrast, I have bracketed the issue of whether these structural formulations are 'accurate' representations of the world, and treated them in terms of their performativity. That is to say, I have looked at how such terms circulate in the attempted 'making' of particular (future) worlds. Moreover, we have addressed, albeit in passing, the question of what sort of world is 'made' in the cross-circulations of these contrasting grand societal formulations.

Over the course of this book, I have made several proposals about the way that the interrelations of technoscience and everyday life might be theorized. For instance, I have tentatively suggested that the complexities of technoscientifically mediated surveillance in everyday life might be rethought through the motif of the taleidoscope and the figure of the 'taleidoscopticon'; I have argued that one way of attending to the shifting dynamics of everyday politics as they are engaged with technoscientific matters is through the construct of 'ethno-epistemic assemblages'; and I have recommended that everyday engagements with technoscience can be imagined in terms of an array of 'co-present' spatialities and temporalities, and, in the case of the former, have mused that some spatialities can be related to one another through new prepositions such as 'frottage'. Overarching these interventions is an analytic ethos that is orientated towards such concerns as relationality, emergence, heterogeneity, multiplicity and complexity.

It will not have escaped notice that the relation between the various examples I have used, and my thematic constructs of the body, citizenship, society, spatiality and temporality, can be seen to be rather arbitrary. The 'railway station queue' that was deployed in illustration of the complex enactments of temporality (or, more accurately, temporal performativities) can just as easily be used in the exemplification of the impure purification of society, or of the 'looking down' into the local to unravel the 'global', or of the dynamics of everyday micropolitics, or of the open/closed-ness of the body. The trajectories of Post-it® Notes could have served not only in the tracing of the performances of particular sorts of society, but also as moments in the making of ethno-epistemic assemblages, or as means to the exploration of new prepositions with which to capture complex mundane spatialities, or as loci in the enactment of multiple bodies, or as instances in the conjoint ordering and disordering of everyday temporalities in the office. This interchangeability is, of course, a partial upshot of what have become self-evident, conceptual linkages between these thematic constructs: bodies, politics, societies, times and spaces are nowadays co-implicated, not least in the doings of everyday life. Indeed, below, I will draw again on a particular technoscientific artefact – velcro – to illustrate one way in which these thematic constructs can be further related to one another. What all this reflects is a conceptual assemblage (or ensemble) – or, better still, an analytic sensibility – for thinking about how such artefacts operate multiply in the enactment of the corporeal, political, societal, spatial and temporal complexities of everyday life. If this implies woolliness of thought (and method), this seems to me to be well suited to the woolliness of the everyday world.

To reiterate, there is no aspiration here to sculpt an all-encompassing theoretical framework for the study of technoscience and everyday life.

A critical thought or two on 'critique'

Despite a level of analysis orientated towards the micro, I have nevertheless attempted to engage in 'critique', albeit without making any strong claims about (macro) structures of power differentials, inequality, exclusion and the like. This is because, as is often argued from the micro perspective, the key thing is to investigate how such 'structures' emerge through the micro processes of everyday life, including those processes that serve to generate accounts about such structures (and, indeed, everyday life itself). Putting together critique and micro analysis might yield Crook's (1998) minotaur, but such monsters are, as I hope I have amply illustrated, also the stuff of the world.

Let me consider the double aspect of the present monstrous version of critique.[2] First, throughout this book, I have engaged with the relations between technoscience and everyday life in terms of the dynamics of contestation (and collaboration). Critique of technoscience-and-everyday life,[3] as we might now call it, is already present within the doings of technoscience-and-everyday life. Versions of bodies, citizens, societies, times and spaces are advocated and contested as part of the very fabric of technoscience-and-everyday life. The present contribution is simply another critical voice that enters into the melee, so to speak, out of its own peculiar (contestable) everyday life (academic). To the extent that the critiques contained in this volume have indulged in abstraction towards the macro, we can claim common sensical warrantability: abstraction towards the macro is part and parcel of everyday argumentation and 'ordinary' critique (as I hope to have shown in relation to, for example, people's everyday rhetorical use of 'society'). In other words, the present critique does not 'stand above' the multitude of critiques that circulate in the making of technoscience-and-everyday life; or, rather, such 'standing above' is an aspiration and claim shared by many of those everyday critiques whose circulation inscribes the mundane. On this account, the extraordinariness of critique is its ordinariness, and vice versa.

The second aspect of the present version of critique concerns a rather different sort of monstrosity. Much of the critical commentary in this book has been directed at exposing, in some small measure, a range of 'unfairnesses' enacted within technoscience-and-everyday life. We have also documented how groups and individuals struggle against and resist such 'unfairnesses'. However, this 'humanist critique' sits side by side with a more 'heterogeneous critique'; after all, such individuals and groups emerge out of the complex arrays of sociotechnical assemblages. This 'heterogeneous critique' amounts, in part, to an always contingent 'laying bare' of the patternings of orderings/disorderings that comprise everyday life as partially mediated by the complex circulations of technoscience. However, 'laying bare' amounts to little without the material/semiotic (and technoscientifically mediated) circulations of the *present text* that *possibly* might affect these patternings – serving in the disruption of some orderings and in the reinforcement of others – of everyday life in some small way. Yet, as we have been at pains to point out, the circulation of artefacts such as this book, whatever the pretensions of their 'scripts' (Akrich 1992), entails the reshaping of those artefacts along 'their' trajectory. As we have repeatedly seen, such artefacts change their 'identity' by virtue of the associations they form in the process of ordering and disordering. But such identity shifts or shimmerings also arise partly because it is not artefacts as singularities that circulate, but a particular (or particulate) manifestation of this or that sociotechnical assemblage that is nothing if not complex, multivalent, etc. (i.e. shifting and shimmering).

Three implications can be drawn from this sketch of 'heterogeneous critique'. The first is that it is about affecting patternings of ordering and disordering (though this certainly does not preclude 'championing' this or that disadvantaged human, or indeed nonhuman, grouping). Second, the artefact that does the 'heterogeneous critiquing' itself gets caught up in, and partially connected with and, to a significant degree, transformed by, the process of being 'effective'. As a corollary, such a critique is marked by unintended consequences – or rather, *overtly* marked by unintended consequences; it is, in this respect, 'experimental' (Latour 2004a). Third, 'heterogeneous critique' is not mediated by a singular artefact because such artefacts are particulate manifestations of complex, heterogeneous assemblages. Accordingly, the critique enacted in and by the present volume is simply one exemplification (or 'singularization') of the critical interventions of a dynamic heterogeneous assemblage.

Here we have touched upon the thorny issue of identity (which this chapter is ostensibly about). As the foregoing suggests, technoscience is neither 'itself' uniform or unitary, nor are its artefacts (whether these be individual(ized) technologies or projections of future states of affairs) simple or singular. For instance, 'xenotransplantation' comprises a multitude of relations, entities, arguments, models, risks and regulations, and so on and so forth. Or, rather, xenotransplantation is manifested or expressed, or – best of all and to draw on Whitehead – concresced in different ways depending on the sorts of relations, entities, arguments, models, risks and regulations, and so on and so forth, that come together. Thus, xenotransplantation is a logical and relatively minor biotechnological advance, or a major revision of the human condition; it is a travesty for animal rights or, at once, the realization and dilution of patients' treatment opportunities; it is a domain of regulatory rigour or a further spur to expert migration, or another expression of technoscientific hubris. Related to these multiple identities of xenotransplantation are, of course, multiple human identities – as individual bodies and species capacities, as patients caught up in the processes of technoscientific 'innovation' and concerned citizens, as scientists, regulators and more or less disenfranchised publics; as prospective and privileged consumers of this technology, and as sceptics about the temporal and spatial promises associated with xenotransplantation. The list goes on ...

In the next section, I try to impose, with all the usual provisos of modesty, tentativeness and woolliness in place, some order on these complex, dynamic and multiple doings of identity in technoscience-and-everyday life. This I do by reconsidering velcro, which serves as an example through which to identify (*sic*) how identity in technoscience-and-everyday life might be wrought.

Re-thinking (with) velcro

Let me begin with an example or two.

When my daughter was two she got her first pair of sandals with velcro fastenings at the toes and ankle. She was enormously proud, as I was, that she could put her shoes on all by herself – they might not have been on the right feet, but they were fixed securely and she did it all by herself. My daughter also wears a cycling helmet. It is attached to her with a clip-on strap, and inside are little cushions stuck to the helmet via velcro pads that serve to make the helmet fit as snugly as possible. She has very fine hair and this routinely gets caught up in the velcro. Many unhappy moments have been spent unpicking her hair from the Velcro as she screams and squirms.

In Chapter 1, we encountered the Edinburgh Bus and Coach Watchdog and its report which accused a local bus company of using inadequate velcro strips to attach seat cushions to seat frames. The problem was that passengers, using the seats for support as they moved through the bus, could, if the cushions became detached, be thrown forward and injured. In contrast to this specific problem, we also came across a more generic issue. To recall, the Christian Reformed Church of North America argued that the ease associated with the use of velcro deprives children of opportunities for everyday problem solving and reduces their capacities to 'learn how to learn'. As hinted in Chapter 1, this account can be viewed as grounded in a principle (one might say caricature) of the 'utter convenience' of velcro. Nevertheless, it is a narrative concerned with the 'good life': shoelaces are a means to better realizing human potential. This caricature also informs the way that velcro has been appropriated as a way of describing the potential use of carbon nanotubes to produce molecular adhesives – nano-velcro – of unprecedented strength. Nano-velcro also looks to the future, but it is an instrumental future in which we get to the same place with better technoscience (i.e. stick stuff together ever more effectively).

Each of these examples reflects a particular identity for velcro-and-humans. In relation to the process of extricating my daughter's hair from velcro, we encounter an example of the micro-routines of everyday life. In terms of the micro approach enacted in this book, everyday life is shaped by these encounters with mundane technoscientific artefacts. Bodies are abled and disabled, embodied knowledge is at once mediated by these artefacts and, in turn, mediates these artefacts in a series of local, practical collectively co-ordinated micro-innovations that allow those artefacts to 'work successfully'. Further, these mutual accommodations, mediations and translations can be seen as indices of co(a)gency: 'ultimately', what is 'working successfully' is artefact-and-body together.

Turning to the case of velcro, bus cushions and the protests by the Edinburgh Bus and Coach Watchdog, we might link this up with De Certeau's perspective. In the spaces of a bus, there are particular specified forms of comportment – most relevantly the use of the hand rails provided. The use of the cushions is what De Certeau would call a tactic or ruse – one of those minuscule practices that resists, for a fleeting moment, the exercise of sur-veillance within what he would presumably see as the strategic space of the bus. In contrast, the report itself is arguably an example of what Lefebvre calls 'the bureaucratic society of controlled consumption', in which a bureaucratic form of rationality operates upon consumption within everyday life. In other words, people bureaucratically oversee their own forms of consumption – here they write a report about improving the conditions of consumption of a public transport service. On the one hand, there is a sort of minimal usurp-ation of strategic space – on the other, a resistance that turns out to be about doing consumption more efficiently. In the present context, the circulations of the report can also be seen as an exemplification of the operations of the taleidoscopticon in so far as the bus company itself is placed under surveil-lance. Thus a 'citizenly' intervention is thoroughly intertwined with 'con-sumerly' action (just as, conversely, the 'consumerly' practice of choosing to travel by bus, as opposed to car, might be a citizenly act).

These analyses are not altogether convincing and that is because they reduce the complexity of the everyday. Everyday life here is generally im-poverished by being simplified around motifs of panopticism or 'the bu-reaucratic society of controlled consumption'; these 'oppressive' versions of everyday life are then juxtaposed to little bouts of compromised resistance (not least, those that can be drawn into the motif of the 'taleidoscopticon'). Instead, we might work with a more processual model of the everyday, and attune ourselves to the disordering and ordering processes that comprise everyday life. As such, citizenly and consumerly enactments are interventions in, and expressions of, the dynamic patternings of sociotechnical assemblages.

When we reconsider the lessons of the laces (as opposed to the violations of velcro), we venture a little closer towards a vision of a better future. To be sure, this is hardly a Lefebvrian 'moment of presence' that punctures the taken-for-grantedness of the everyday with an insight into the totality of society and the realization of 'total man' (*sic*). Nevertheless, it does engage with a concern for the impoverishment of human potential in so far as lace-tying implies a humanist and ends-orientated view of a possible future in which people will relearn how to learn. By contrast, the revels of nano-velcro are highly technologistic and means-orientated – people will be able to pursue their practical goals (in much the same way as they do now) but with the aid of fantastically more adhesive adhesives. In some ways, both these uses of velcro might be regarded as 'spectacular' (in the situationist sense), because of the ways in which they neglect the vicissitudes of velcro, and draw on velcro as an abstracted principle, or caricature.

On another level, these allegorical and metaphorical uses of velcro are partly concerned with constituting the present in order to effect a particular future. In the case of laces, the future is found in an idyllic past. In the case of nano-velcro, the future rests on the imminent prospect of invention and innovation. In both cases, however, there is an engagement, in one way or another, with making the present so that the future is made. Each account tacitly offers a way of marshalling people, technoscientific artefacts, texts and relations in the here and now in order to accomplish a particular, desired end – that is, the future.

We are now in a position to abstract from the foregoing some provisional, though hardly exhaustive, implications for 'identity'.

- We might say that the velcro-and-helmet episode relates to the intersubjective and intercorporeal processes whereby adjustments are made to local technoscientific configurations. Identity here is enacted in the local process of mutual translations between technologies and bodies. Here identity emerges in the context of what we might call *micro-mutual innovation*.
- We might say that the velcro-and-bus-cushion incidents and report entail the enactments of political actors – citizenly-consumerly identities that are performed in relation to bus companies and regulatory authorities. So, in this case, identity is performed through what we might call *citizenly-consumerly practices*. However, such practices can also be said to be concerned with the sort of identity one has as someone more or less 'integrated into' an everyday sociotechnical assemblage (such as a local transport system). Or, rather, citizenly-consumerly practices are heterogeneous interventions in, and expressions of, such sociotechnical assemblages – assemblages mediated by, among other things, co(a)gents comprising combinations of people, velcro, seats, poles, PCs, printers (for the production

of critical reports), and so on. Here, identity is embroiled in the *re-patterning co(a)gencies* that characterize the ordering/disordering dynamics of sociotechnical assemblages.

- We might say that both velcro-and-laces and nano-velcro texts point to prospective identities – to matters of hope and futurity in which more or less extravagant potentials can be pursued. However, as repeatedly noted, projections of the future are enactments in the present: they are concerned with the (re)making of here and now (and also the past, needless to say) that will, it is hoped, make a where and when. Here identity is enacted in both the doing of, and the representations entailed in, what we called in Chapter 7 *temporal performativities*.

Predictably, it turns out that the associations I have drawn between illustrative case and identity are more complex.

- The velcro-and-helmet incident could just as easily serve as the grounds for citizenly-consumerly practices in relation to the manufacturers, or regulatory authorities (e.g. British Safety Standards); or be a moment in the repatterning co(a)gencies that make up local traffic flows; or be attached to environmental politics whose temporal performativities paint a world in which environmental sustainability is attained.
- We would not be surprised to find that the velcro-and-bus-cushion incidents often entail micro-mutual innovation: many people simply work around the exigencies of such an 'inconvenience' as dangerous seat cushions with the inventive help of fellow travellers and their own creative use of whatever other artefacts come to hand. We would not be surprised to find that such citizenly-consumerly interventions contain the trace of possible transport futures where, say, users are routinely and properly consulted in the bus design process. Indeed, the co(a)gential repatternings of the report can be said to involve temporal performativities through which future relations among bus companies, regulators, users, manufacturers are assigned, and thus present relations reconfigured.
- In the case of the velcro-and-laces texts, the sermonizing on the 'good' of laces presupposes a fruitful interactivity between laces and person, or productive micro-mutual innovation. Moreover, this text can be read as citizenly-consumerly intervention in the manufacture of children's shoes, and more broadly in parental responsibility; and in its circulation, such a text is a repatterning co(a)gent that reflects and affects the dynamics of sociotechnical assemblages in which, for example, the everyday lives of children are ordered and disordered.

The general point here is that these four ways of doing identity in techno-science-and-everyday life – micro-mutual innovation, citizenly-consumerly practices, repatterning co(a)gencies, and temporal performativities – are co-present. To make things more complex still, those instances that have focused on velcro, of course, cohabit with myriad other more or less mundane technoscientific artefacts and technoscientific trajectories: cat's eyes and cycle paths, hip replacements and pneumatic braking systems, information and communication technologies and capacity-building initiatives for nano-technology. In the midst of such miscellany, these doings of identity that I have put forward can only be blunt heuristic suggestions that, for all their terminological otherworldliness, might serve as prompts for thinking about the complex identities that populate and mediate technoscience-and-everyday life. Having pointed yet again to complexity, we must not forget that there are concurrent attempts at technoscientific simplification. The multifariousness of doings in technoscience-and-everyday life is accountable in terms of singular, epochal identity narratives: this is the age of genetic selves, or somatic selves, or posthuman selves, or surveilled selves, to mention but a few (Butler 1999). And yet, as showed in Chapter 5, these numerous moves to narrate (more or less exclusively, more or less encompassingly) the present era (whether in terms of society or in terms of identity) contribute to its very complexity.

Technoscience, tools and trajectories

Velcro is, in its multifariousness and heterogeneity, a series of technoscientific artefacts, a range of technoscientific stories, and an encapsulation and manifestation of several sociotechnical assemblages. On this score, there is nothing unusual about velcro. Though it, and a multitude of other artefacts, can – perhaps with the help of some of the ideas outlined in the foregoing – serve as a tool that helps us trace a trajectory through the corporeal (extra)-ordinarinesses, the political and social contestabilities, the temporal and spatial promises that make up the complex byways of everyday life. In the end, we might view a technoscientific artefact like velcro as a tool for pulling out and weaving together a few hopeful threads from the woolliness of the everyday.

Notes

1. In light of this dynamism, we should bear in mind that terms like 'assemblage' (or 'ensemble') are merely suggestive terms that go only a certain way to giving expression to the – or, rather, performing – processuality and complexity of the relations that make up technoscience and everyday life.

2. There is much to be unpacked about the relation of the present discussion of critique to others, notably those within the critical theoretical tradition and its dichotomy of immanent and utopian critiques (e.g. Geuss 1981). Suffice it to say that this book has much in common with the critique embodied in Mick Billig's approach to rhetorical analysis (e.g. Billig, Condor, Edwards, Gane, Middleton and Radley 1988).

3. The hyphenation of the title of this book is meant to mark, at last, the complex inseparabilities – empirically and theoretically – of technoscience and everyday life.

References

Abelson, J., Forest, P.-G., Eyles, J., Smith, P., Martin, E. and Gauvibin, F.-P. (2003) Deliberations about deliberation: issues in the design and evaluation of public consultation processes. *Social Science and Medicine*, 57, 239–251.

Adam, B. (1998) *Timescapes of Modernity*. London: Routledge.

Adam, B., Beck, U. and Van Loon, J (eds) (2000) *The Risk Society and Beyond*. London: Sage.

Akrich, M. (1992) The de-scription of technical objects. In W.E. Bijker and J. Law, (eds) *Shaping Technology/Building Society*. Cambridge, MA.: MIT Press, 205–224.

Akrich, M. and Latour, B. (1992) A summary of a convenient vocabulary for the semiotics of human and nonhuman assemblies. In W.E. Bijker and J. Law (eds) *Shaping Technology/Building Society*. Cambridge, MA: MIT Press, 259–263.

Amann, K. and Knorr Cetina, K. (1990) The fixation of (visual) evidence. In M. Lynch and S. Woolgar (eds) *Representation in Scientific Practice*. Cambridge, MA: MIT Press, 85–121.

Amin, A. and Thrift, N. (2002) *Cities: Reimagining the Urban*. Cambridge: Polity.

Ansell Pearson, K. and Mullarkey, J. (2002) Introduction. In K. Ansell Pearson and J. Mullarkey (eds) *Henri Bergson: Key Writings*. London: Continuum.

Arksey, H. (1998) *RSI and the Experts: The Construction of Medical Knowledge*. London: UCL Press.

Aronowitz, S., Martinsons, B. and Menser, M. (eds) (1996) *Technoscience and Cyberculture*. New York, NY: Routledge.

Attfield, J. (2000) *Wild Things: The Material Culture of Everyday Life*. Oxford: Berg.

Ball, K. (2003) Editorial: The labours of surveillance. *Surveillance and Society*, 1, 125–137, at http://www.surveillance-and-society.org.

Ball, M. (2000) The visual availability and local organization of public surveillance systems: the promotion of social order in public spaces. *Sociological Research Online*, 5, at http://www.socresonline.org.uk/5/1/ball.html.

Barbalet, J.M. (2001) *Emotion, Social Theory and Social Structure*. Cambridge: Cambridge University Press.

Barry, A. (2001) *Political Machines*. London: Athlone.

Beardsworth, A. and Keil, T. (1997) *Sociology on the Menu: An Invitation to the Study of Food and Society*. London: Routledge.

Beck, U. (1992) *The Risk Society*. London: Sage.

Beck, U. and Beck-Gernsheim, E. (1995) *The Normal Chaos of Love*. Cambridge: Polity Press.

Beck, U. and Beck-Gernsheim, E. (2001) *Individualization*. London: Sage.

Beck, U., Giddens, A. and Lash, S. (1994) *Reflexive Modernization: Politics, Tradition and Aesthetics in the Modern Social Order*. Cambridge: Polity.

Beder, S. (1999) Public participation or public relations. In B. Martin (ed.) *Technology and Public Participation*. Wollongong, Australia: Science and Technology Studies, University of Wollongong, 169–292.

Bennett, T. (2004) The invention of the modern cultural fact: a critique of the critique of everyday life. In E. Silva and T. Bennett (eds) *Contemporary Culture and Everyday Life*. Durham: sociologypress.

Bennett, T. and Watson, D. (eds) (2002) *Understanding Everyday Life*. Oxford: Blackwell.

Berger, P.L. and Luckmann, T. (1966) *The Social Construction of Reality*. Harmondsworth: Penguin.

Bijker, W.E. (1995) *Of Bicycles, Bakelite and Bulbs: Toward a Theory of Sociotechnical Change*. Cambridge, MA: MIT Press.

Billig, M., Condor, S., Edwards, D., Gane, M., Middleton, D. and Radley, A. (1988) *Ideological Dilemmas*. London: Sage.

Birke, L. (2000) *Feminism and the Biological Body*. Piscataway, NJ: Rutgers University Press.

Birke, L., Arluke, A. and Michael, M. (forthcoming) *The Sacrifice: How Scientific Experiments Transform Animals and People*. Ashland, OH: Purdue University Press.

Bittman, M., Rice, J.M. and Wajcman, J. (2004) Appliances and their impact: the ownership of domestic technology and time spent on household work. *British Journal of Sociology*, 55, 401–422.

Bourdieu, P. (1984) *Distinction: A Social Critique of the Judgement of Taste*. London: Routledge & Kegan Paul.

Bovone, L. (1989) Theories of everyday life. *Current Sociology*, 31, 41–59.

Bowers, J. and Iwi, K. (1993) The discursive construction of society. *Discourse and Society*, 4, 357–393.

Bowker, G.C. and Star, S.L. (1999) *Sorting Things Out: Classification and its Consequences*. Cambridge, MA: MIT Press.

Boyne, R. (2000) Post-panopticism. *Economy and Society*, 29, 285–307.

Brannigan, A. (1981) *Social Bases of Scientific Discovery*. New York, NY: Cambridge University Press.

Braudel, F. (1981) *The Structures of Everyday Life: Civilization and capitalism 15th–18th Century, Volume 1*. London: Phoenix Press.

Brown, B. (2004) The order of service: the practical management of customer interaction. *Sociological Research Online*, 9, at http://www.socresonline.org.uk/9/4/brown.html.

Brown, N. (1998) *Ordering Hope*. Unpublished PhD Thesis, School of Independent Studies, Lancaster University.

Brown, N. (2000) The breakthrough motif. In N. Brown, B. Rappert and A. Webster (eds) *Contested Futures*. Aldershot: Ashgate.

Brown, N. and Michael, M. (2001) Switching between science and culture in transpecies transplantation. *Science, Technology and Human Values*, 26, 3–22.

Brown, N. and Michael, M. (2002) From authority to authenticity: the changing governance of biotechnology. *Health, Risk and Society*, 4(3), 259–272.

Brown, N. and Webster, A. (2004) *New Medical Technologies and Society: Reordering Life*. Cambridge: Polity.

Brown, S.D. and Lightfoot, G. (2002) Presence, absence and accountability: e-mail and the mediation of organizational memory. In S. Woolgar (ed.) *Virtual Society? Technology, Cyberbole, Reality*. Oxford: Oxford University Press, 209–229.

Bull, M. (2000) *Sounding Out the City*. Oxford: Berg.

Burkitt, I. (1999) *Bodies of Thought: Embodiment, Identity and Modernity*. London: Sage.

Butler, J. (1999) *Gender Trouble*. New York, NY: Routledge.

Callon, M. (1986a) The sociology of an actor-network: the case of the electric vehicle. In M. Callon, J. Law and A. Rip (eds) *Mapping the Dynamics of Science and Technology*. London: Macmillan, 19–34.

Callon, M. (1986b) Some elements in a sociology of translation: domestication of the scallops and fishermen of St Brieuc Bay. In J. Law (ed.) *Power, Action and Belief*. London: Routledge & Kegan Paul, 196–233.

Callon, M. and Latour, B. (1981) Unscrewing the big Leviathan. In K.D. Knorr Cetina and M. Mulkay (eds) *Advances in Social Theory and Methodology*. London: Routledge & Kegan Paul, 275–303.

Callon, M. and Law, J. (2004) Guest editorial. *Environment and Planning D: Society and Space*, 22, 3–11.

Canguilhem, G. (1994) The normal and the pathological. In F. Delaporte (ed.) *A Vital Rationalist: Selected Writings from Georges Canguilhem*. New York, NY: Zone Books, 321–350.

Carlin, A. (2003) Observation and membership categorization: recognizing 'normal appearances' in public space, at http://mundanebehavior.org/issues/v4n1/carlin.htm (accessed 14 August 2005).

Carter, S. and Michael, M. (2004) Here comes the sun: shedding light on the cultural body. In J. Ahmed and H. Thomas (eds) *Cultural Bodies: Ethnography and Theory*. New York, NY: Blackwell, 260–282.

Carter, S., Green, J., Wardell, C. and Thorogood, N. (2005) The domestication of everyday health technology: a case study of electric toothbrushes. Unpublished manuscript.

Castells, M. (2001) *The Internet Galaxy: Reflections on the Internet, Business, and Society*. Oxford: Oxford University Press.

Chaney, D. (1993) *Fictions of Collective Life*. London: Routledge.

Cohen, A.P. (1985) *The Symbolic Construction of Community*. Chichester: Ellis Harwood.

Cohen, S. and Taylor, L. (1992) *Escape Attempts: The Theory and Practice of Resistance to Everyday Life* (2nd edn). London: Routledge.

Collins, H.M. (1985) *Changing Order.* London: Sage.

Collins, H.M. (1988) Public experiments and displays of virtuosity. *Social Studies of Science,* 18, 725–748.

Conrad, P. and Gabe, J. (1999) Introduction. In P. Conrad and J. Gabe (eds) *Sociological Perspectives on the New Genetics.* Oxford: Blackwell, 1–12.

Cooper, G., Green, N., Murtagh, G.M. and Harper, R. (2002) Mobile society? Technology, distance and presence. In S. Woolgar (ed.) *Virtual Society? Technology, Cyberbole, Reality.* Oxford: Oxford University Press, 286–301.

Cowan, R.S. (1985) The industrial revolution in the home. In D. MacKenzie and J. Wajcman (eds) *The Social Shaping of Technology.* Buckingham: Open University Press.

Crook, S. (1998) Minotaurs and other monsters: 'everyday life' in recent social theory. *Sociology,* 32, 523–540.

Crook, S., Pakulski, J. and Walters, M. (1992) *Postmodernization.* London: Sage.

Cwerner, S.B. and Metcalfe, A. (2003) Storage and clutter: discourse and practices of order in the domestic world. *Journal of Design History,* 16, 229–239.

Dant, T. (1999) *Material Culture in the Modern World.* Buckingham: Open University Press.

Dant, T. (2004) The driver-car. *Theory, Culture and Society,* 21, 61–79.

Dean, M. (1999) *Governmentality: Power and Rule in Modern Society.* London: Sage.

Debord, G. (1983) *Society of the Spectacle.* Detroit, MN: Black and Red.

De Certeau, M. (1984) *The Practice of Everyday Life.* Berkeley, CA: University of California Press.

De Certeau, M., Giard, L. and Matol, P. (1998) *The Practice of Everyday Life: Living and Cooking, Vol. 2.* Minneapolis, MN: University of Minnesota Press.

Deleuze, G. and Guattari, F. (1988) *A Thousand Plateaus: Capitalism and Schizophrenia.* London: Athlone Press.

Dickens, B.M. (2002) Can sex selection be ethically tolerated? *Journal of Medical Ethics,* 28, 335–336.

Dingwall, R. (2001) Contemporary legends, rumours and collective behaviour: some neglected resources for medical sociology? *Sociology of Health and Illness,* 23(2), 180–202.

Douglas, M. (1966) *Purity and Danger.* London: Ark.

Durant, J.R., Evans, G.A. and Thomas, G.P. (1989) The public understanding of science. *Nature,* 340 (6 July), 11–14.

Edensor, T. (2000) Walking in the British countryside: reflexivity, embodied practices and ways to escape. *Body and Society,* 6, 81–106.

Elam, M. and Bertilsson, M. (2003) Consuming, engaging and confronting science: the emerging dimensions of scientific citizenship. *European Journal of Social Theory,* 6, 233–251.

Elias, N. (1939/1994) *The Civilizing Process.* Oxford: Blackwell.

Epstein, S. (1996) *Impure Science: AIDS Activism and the Politics of Science.* Berkeley, CA: University of California Press.

Epstein, S. (2000) Democracy, expertise and AIDS treatment activism. In D.L. Kleinman (ed.) *Science, Technology and Democracy*. Albany, NY: State University of New York Press, 15–32.

Fairclough, N. (1992) *Discourse and Social Change*. Cambridge: Polity.

Featherstone, M. (1991) *Consumer Culture and Postmodernism*. London: Sage.

Featherstone, M. (1992) The heroic life and everyday life. *Theory, Culture and Society*, 9, 159–12.

Featherstone, M. (2004) Automobilities: an introduction. *Theory, Culture & Society*, 21, 1–24.

Feenberg, A. (1999) *Questioning Technology*. London: Routledge.

Felski, R. (1999–2000) The invention of everyday life. *Transformation*, 39, 13–31.

Felt, U. (2003) *Optimising Public Understanding of Science and Technology in Europe. Final Report to the European Commission*, at http://www.univie.ac.at/virusss/OPUSReport.

Fischer, F. (2000) *Citizens, Experts and the Environment: The Politics of Local Knowledge*. Durham, NC: Duke University Press.

Fleck, L. (1979) *Genesis and Development of a Scientific Fact*. Chicago, IL: Chicago University Press.

Foucault, M. (1979) *Discipline and Punish*. Harmondsworth: Penguin.

Foucault, M. (2003) *The Birth of the Clinic*. London: Routledge.

Franklin, A. (1999) *Animals and Modern Cultures: A Sociology of Human–Animal Relations in Modernity*. London: Sage.

Franklin, S., Lury, C. and Stacey, J. (2000) *Global Nature, Global Culture*. London: Sage.

Frissen, V. (1995) Gender is calling: some reflections on past, present and future uses of the telephone. In K. Grint and R. Gill (eds) *The Gender–Technology Relation*. London: Taylor & Francis, 79–94.

Fukuyama, F. (2002) *Our Posthuman Future*. New York, NY: Profile Books.

Funtowicz, S.O. and Ravetz, J. (1993) Science for the post-normal age. *Futures*, 25/7, 735–755.

Gabriel, Y. and Lang, T. (1995) *The Unmanageable Consumer: Contemporary Consumption and its Fragmentation*. London: Sage.

Gardiner, M. (2000) *Critiques of Everyday Life*. London and New York: Routledge.

Garfinkel, H. (1967) *Studies in Ethnomethodology*. Cambridge: Polity Press.

Garfinkel, H. and Livingston, E. (2003) Phenomenal field properties of the order of service in formated queues and their neglected standing in the current situation of inquiry. *Visual Studies*, 18, 21–28.

Garforth, L. (2005) No expectations: utopian theory after the future. Paper presented at Researching Expectations in Medicine, Technology and Science – Social Theory and Methodology. Expectations Network Meeting. University of York.

Gershuny, J. (2004) Domestic equipment does not increase domestic work: a response to Bittman, Rice and Wacjman. *British Journal of Sociology*, 55, 425–431.

Geuss, R. (1981) *The Idea of a Critical Theory*. Cambridge: Cambridge University Press.

Gibson, E.E. (1979) *The Ecological Approach to Visual Perception*. Boston, MA: Houghton Mifflin.

Giddens, A. (1984) *The Constitution of Society*. Berkeley, CA: University of California Press.

Giddens, A. (1990) *The Consequences of Modernity*. Stanford, CA: Stanford University Press.

Giddens, A. (1991) *Modernity and Self-identity*. Cambridge: Polity Press.

Giddens, A. (1992) *The Transformation of Intimacy*. Cambridge: Polity Press.

Giddens, A. (1998) *The Third Way: The Renewal of Social Democracy*. Cambridge: Polity Press.

Gieryn, T.F. (1999) *Cultural Boundaries of Science: Credibility on the Line*. Chicago: University of Chicago Press.

Goffman, E. (1959) The *Presentation of Self in Everyday Life*. Harmondsworth: Penguin.

Goffman, E. (1961) *Asylums*. Harmondsworth: Penguin.

Goffman, E. (1964) *Stigma*. Harmondsworth: Penguin.

Gouldner, A. (1975) Sociology and everyday life. In L.A. Coser (ed.) *The Idea of Social Structure*. New York, NY: Harcourt, Brace, Jovanovich, 417–432.

Graham, E.L. (2002) *Representations of the Post/Human: Monsters, Aliens and Others*. Manchester: Manchester University Press.

Gregory, J. and Miller, S. (1998) *Science in Public: Communication, Culture and Credibility*. New York, NY: Plenum.

Gubrium, J.F. and Holstein, J.A. (eds) (2000) *Aging and Everyday Life*. Malden, MA: Blackwell.

Gullo, A., Lassiter, U. and Wolch, J. (1998) The cougar's tale. In J. Wolch and J. Emel (eds) *Animal Geographies: Place, Politics and Identity in the Nature–Culture Borderlands*. London: Verso, 139–161.

Guthman, J. (2003) Fast food/organic food: reflexive tastes and the making of 'yuppie chow'. *Social and Cultural Geography*, 4, 45–58.

Habermas, J. (1987) *The Theory of Communicative Action, Vol. 2*. Cambridge: Polity.

Hacking, I. (1986) Making up people. In T.C. Heller, M. Sosna and D.E. Wellberg (eds) *Reconstructing Individualism*. Stanford, CA: Stanford University Press, 222–236.

Hagendijk, R. and Kallerud, E. (2003) Changing conceptions and practices of governance in science and technology in Europe: a framework for analysis. *Discussion Paper Two: Science, Technology and Governance in Europe*.

Haraway, D. (1991) *Simians, Cyborgs and Nature*. London: Free Association Books.

Haraway, D. (1997) *Modest_Witness@Second_Millenium.FemaleMan.Meets_Onco Mouse: Feminism and Technoscience*. London: Routledge.

Harvey, D. (1989) *The Condition of Postmodernity*. Oxford: Blackwell.

Hausmann, E. (2002) *Media Representations of Euthanasia*. Unpublished PhD thesis, Goldsmiths College, University of London.

Hayles, N.K. (1999) *How We Became Posthuman: Virtual Bodies in Cybernetics, Literature and Informatics*. Chicago, IL: University of Chicago Press.

Haynes, R.D. (1994) *From Faust to Strangelove: Representations of the Scientist in Western Literature*. Baltimore, MD: Johns Hopkins University Press.

Hedgecoe, A. (2001) Ethical boundary work: geneticization, philosophy and the social sciences. *Medicine, Health Care and Philosophy*, 4, 305–309.

Held, D. (1987) *Models of Democracy*. Cambridge: Polity.

Heller, A. (1984) *Everyday Life*. London: Routledge.

Heritage, J. (1984) *Garfinkel and Ethnomethodology*. Cambridge: Polity Press.

Hess, D.J. (1995) *Science and Technology in a Multicultural World*. Columbia, NY: Columbia University Press.

Highmore, B. (2002) *Everyday Life and Cultural Theory: An Introduction*. London: Routledge.

Hill, A. and Michael, M. (1998) Engineering acceptance: representations of 'the public' in debates on biotechnologies. In P. Wheale, R. von Schomberg and P. Glasner (eds) *The Social Management of Genetic Engineering*. London: Avebury, 201–217.

Hillman, D. and Gibbs, D. (1998) *Century Makers*. London: Weidenfeld & Nicolson.

Hislop, J. and Arber, S. (2003) Sleepers wake! The gendered nature of sleep disruption among mid-life women. *Sociology*, 37(4), 695–711.

Honore, C. (2005) *In Praise of Slow: How a Worldwide Movement is Challenging the Cult of Speed*. London: Orion.

House of Lords Select Committee on Science and Technology (2000) *Third Report: Science and Society. HL Paper 38*. London: The Stationery Office.

Hughes, R. (2004) *Goya*. London: Vintage.

Ingold, T. (1993) The temporality of the landscape. *World Archeology*, 25, 152–174.

Institute for Prospective Technological Studies (2005) *Biometrics at the Frontiers: Assessing the Impact on Society (EU 21585 EN)*. Brussels: European Commission.

Irwin, A. (1995) *Citizen Science: A Study of People, Expertise and Sustainable Development*. London: Routledge.

Irwin, A. (2001) Constructing the scientific citizen: science and democracy in the biosciences. *Public Understanding of Science*, 10, 1–18.

Irwin, A. and Michael, M. (2003) *Science, Social Theory and Public Knowledge*. Maidenhead. Berks: Open University Press/McGraw-Hill.

Irwin, A. and Wynne, B. (eds) (1996) *Misunderstanding Science? The Public Reconstruction of Science and Technology*. Cambridge: Cambridge University Press.

Jones, G., McLean, C. and Quattrone, P. (2004) Spacing and timing. *Organization*, 11, 723–741.

Jonsson, B. (1999) *Ten Thoughts about Time: How to Make More of the Time in Your Life*. London: Robinson.

Joss, S. (1999) Introduction: public participation and technology policy – and decision-making – ephemeral phenomenon or lasting change? *Science and Public Policy*, 26, 291–293.

Keat, R., Whiteley, N. and Abercrombie, N. (eds) (1994) *The Authority of the Consumer*. London: Routledge.

Keller, E.F. (1983) *A Feeling for the Organism: Life and Work of Barbara McClintock*. New York, NY: W.H. Freeman & Company.

Keller, E.F. (2001) *The Century of the Gene*. Cambridge, MA: Harvard University Press.

Kellner, D. (no date) Ernst Bloch, utopia and ideology critique, at http://www.uta.edu/english/dab/illumination/kell1.html (accessed 21 July 2005).

Kern, S. (1983) *The Culture of Time and Space, 1880–1918*. London: Weidenfeld & Nicolson.

Kerr, A. and Cunningham-Burley, S. (2000) On ambivalence and risk: reflexive modernity and the new human genetics. *Sociology*, 32, 283–304.

Knabb, K. (ed.) (1981) *Situationist International Anthology*. Berkeley, CA: Bureau of Public Streets.

Knorr Cetina, K. (1988) The micro-social order: towards a reconception. In N.G. Fielding (ed.) *Actions and Structure: Research Methods and Social Theory*. London: Sage, 21–53.

Knorr Cetina, K.D. (1997) Sociality with objects: social relations in postsocial knowledge societies. *Theory, Culture and Society*, 14, 1–30.

Knorr Cetina, K. (1999) *Epistemic Cultures: How the Sciences Make Knowledge*. Cambridge, MA: Harvard University Press.

Lally, E. (2002) *At Home with Computers*. Oxford: Berg.

Lamvik, G.M. (1996) A fairy tale on wheels: the car as a vehicle for meaning within a Norwegian subculture. In M. Lie and K.H. Sorensen (eds) *Making Technology Our Own? Domesticating Technologies into Everyday Life*. Oslo: Scandinavian University Press.

Lash, S. (1988) Discourse or figure? Postmodernism as a 'regime of signification'. *Theory, Culture and Society*, 5, 311–336.

Lash, S. and Urry, J. (1994) *Economies of Signs and Space*. London: Sage.

Lash, S., Quick, A. and Roberts, R. (1998) Introduction: millenniums and catastrophic times. In S. Lash, A. Quick and R. Roberts (eds) *Time and Value*. Oxford: Blackwell, 1–15.

Latour, B. (1987a) *Science in Action: How to Follow Engineers in Society*. Milton Keynes: Open University Press.

Latour, B. (1987b) Enlightenment without critique: a word on Michel Serres' philosophy. In A. Phillips (ed.) *Contemporary French Philosophy*. Cambridge: Cambridge University Press.

Latour, B. (1990) Drawing things together. In M. Lynch and S. Woolgar (eds) *Representations in Scientific Practice*. Cambridge, MA: MIT Press, 19–68.

Latour, B. (1992) Where are the missing masses? A sociology of a few mundane artifacts. In W.E. Bijker and J. Law (eds) *Shaping Technology/Building Society*. Cambridge, MA: MIT Press, 225–258.

Latour, B. (1993) *We have Never been Modern*. Hemel Hempstead: Harvester Wheatsheaf.

Latour, B. (1998) Virtual society: the social science of electronic technologies. Paper presented at the CRICT 10th Anniversary Conference, Brunel University.

Latour, B. (1999) *Pandora's Hope: Essays on the Reality of Science Studies*. Cambridge, MA: Harvard University Press.

Latour, B. (2004a) *Politics of Nature*. Cambridge, MA: Harvard University Press.

Latour, B. (2004b) How to talk about the body? The normative dimension of science studies. *Body and Society*, 10, 205–229.

Latour, B. and Woolgar, S. (1979) *Laboratory Life: The Social Construction of Scientific Facts*. London: Sage.

Laurier, E., Whyte, A. and Buckner, K. (2001) An ethnography of a neighbourhood café: informality, table arrangements and background noise. *Journal of Mundane Behavior*, 2, 195–227.

Law, J. (1987) Technology and heterogeneous engineering: the case of Portuguese expansion. In W.E. Bijker, T.P. Hughes and T. Pinch (eds) *Social Construction of Technological Systems*. Cambridge, MA: MIT Press, 111–134.

Law, J. (1994) *Organizing Modernity*, Oxford: Blackwell.

Law, J. (2004) And if the global were small and noncoherent? Method, complexity, and the baroque. *Environment and Planning D: Society and Space*, 22, 13–26.

Law, J. and Urry, J. (2004) Enacting the social. *Economy and Society*, 33, 390–410.

Lawrence, C. and Shapin, S. (eds) (1998) *Science Incarnate: Historical Embodiments of Natural Knowledge*. Chicago, IL: Chicago University Press.

Layton, D., Jenkins, E., MacGill, S. and Davey, A. (1993) *Inarticulate Science?* Driffield, E. Yorks: Studies in Education Ltd.

Lefebvre, H. (1947) *Critique of Everyday Life*. London: Verso.

Lefebvre, H. (1974) *The Production of Space*. Oxford: Blackwell.

Lefebvre, H. (1984) *Everyday Life in the Modern World*. London: Continuum.

Lemke, T. (2004) Disposition and determinism – genetic diagnostics in risk society. *The Sociological Review*, 52, 550–566.

Lie, M. and Sorensen, K.H. (1996) Making technology our own? Domesticating technologies into everyday life. In M. Lie and K.H. Sorensen (eds) *Making Technology Our Own? Domesticating Technologies into Everyday Life*. Oslo: Scandinavian University Press.

Lippmann, A. (1998) The politics of health: geneticization versus health promotion. In S. Sherwin (ed.) *The Politics of Women's Health: Exploring Agency and Autonomy*. Philadelphia: Temple University Press, 64–82.

Lupton, D. (1999) Monsters in metal cocoons: 'road rage' and cyborg bodies. *Body and Society*, 5, 57–72.

Lury, C. (1996) *Consumer Culture*. Cambridge: Polity.

Lynch, M. (1985) *Art and Artifact in Laboratory Science*. London: Routledge.

Lynch, M. (1993) *Scientific Practice and Ordinary Action: Ethnomethodology and Social Studies of Science*. Cambridge: Cambridge University Press.

Lynch, M. and Woolgar, S. (eds) (1990) *Representation in Scientific Practice*. Cambridge, MA: MIT Press.

Lyon, D. (1994) *The Electronic Eye: The Rise of Surveillance Society*. Cambridge: Polity.

Lyon, D. (2001) *Surveillance Society: Monitoring Everyday Life*. Buckingham: Open University Press.

Lyotard, J.-F. (1984) *The Postmodern Condition: A Report on Knowledge*. Manchester: Manchester University Press.

Mackenzie, A. (2002) *Transductions: Bodies and Machines at Speed*. London: Continuum.

Macnaghten, P. (2003) Embodying the environment in everyday life practices. *The Sociological Review*, 51, 63–84.

Macnaghten, P., Kearnes, M.B. and Wynne, B. (forthcoming) Nanotechnology, governance and public deliberation. What role for the social science? *Science Communication*.

Macnaghten, P. and Urry, J. (1998) *Contested Nature*. London: Sage.

Macnaghten, P. and Urry, J. (eds) (2001) *Bodies of Nature*. London: Sage.

Maffesoli, M. (1989) The sociology of everyday life (epistemological elements). *Current Sociology*, 37, 1–16.

Maffesoli, M. (1996) *The Time of the Tribes*. London: Sage.

Manning, P. (1992) *Erving Goffman and Social Theory*. Cambridge: Polity.

Marlier, E. (1992) Eurobarometer 35.1: opinions of Europeans on biotechnology in 1991. In J. Durant. (ed.) *Biotechnology in Public: A Review of Recent Research*. London: Science Museum, 52–108.

Martin, B. (1999) Conclusion. In B. Martin (ed.) *Technology and Public Participation*. Wallongong, Australia: Science and Technology Studies, University of Wollongong, 249–263.

Martin, E. (1987) *The Woman in the Body*. Buckingham: Open University Press.

Martin, E. (1994) *Flexible Bodies*. Boston, MA: Beacon Press.

Martin, E. (1998) Anthropology and cultural study of science. *Science, Technology and Human Values*, 23, 24–44.

Mason, D., Button, G., Lankshear, G. and Coates, S. (2002) Getting real about surveillance and privacy at work. In S. Woolgar (ed.) *Virtual Society? Technology, Cyberbole, Reality*. Oxford: Oxford University Press, 137–152.

McLuhan, M. (1993) *Understanding Media: The Extensions of Man*. London: Routledge.

Menser, M. and Aronowitz, S. (1996) On cultural studies, science and technology. In S. Aronowitz, B. Martinsons and M. Menser (eds) *Technoscience and Cyberculture*. New York, NY: Routledge, 7–28.

Merriman, P. (2004) Driving places: Marc Augé, non-places, and the geographies of England's M1 motorway. *Theory, Culture & Society*, 21, 145–167.

Merton, R.K. (1973) *The Sociology of Science*. Chicago, IL: Chicago University Press.

Mestrovic, S.G. (1996) *Postemotional Society*. London: Sage.

Michael, M. (1992) Lay discourses of science: science-in-general, science-in-particular and self. *Science, Technology and Human Values*, 17, 313–333.

Michael, M. (1996) Ignoring science: discourses of ignorance in the public understanding of science. In A. Irwin and B. Wynne (eds) *Misunderstanding Science? The Public Reconstruction of Science and Technology.* Cambridge: Cambridge University Press, 105–125.

Michael, M. (1997) Inoculating gadgets against ridicule. *Science as Culture*, 6, 167–193.

Michael, M. (1998a) Between citizen and consumer: multiplying the meanings of the public understanding of science. *Public Understanding of Science*, 7, 313–327.

Michael, M. (1998b) Co(a)gency and cars: the case of road rage. In B. Brenna, J. Law and I. Moser (eds) *Machines, Agency and Desire.* Oslo: TVM, 125–141.

Michael, M. (2000a) *Reconnecting Culture, Technology and Nature: From Society to Heterogeneity.* London: Routledge.

Michael, M. (2000b) These boots are made for walking … mundane technology, the body and human–environment relations. *Body and Society*, 6, 107–126.

Michael, M. (2000c) Futures of the present: from performativity to prehension. In N. Brown, B. Rappert and A. Webster (eds) *Contested Futures.* Aldershot: Ashgate, 21–39.

Michael, M. (2002) Comprehension, apprehension, and prehension: heterogeneity and the public understanding of science. *Science, Technology and Human Values*, 27, 357–370.

Michael, M. (2003) Between the mundane and the exotic: time for a different sociotechnical stuff. *Time and Society*, 12, 127–143.

Michael, M. (2004a) On making data social: heterogeneity in sociological practice. *Qualitative Research*, 4, 5–23.

Michael, M. (2004b) Roadkill: between humans, nonhuman animals, and technologies. *Society and Animals*, 12, 277–298.

Michael, M. and Birke, L. (1994) Animal experimentation: enrolling the core set. *Social Studies of Science*, 24, 81–95.

Michael, M. and Brown, N. (2000) From the representation of publics to the performance of 'lay political science'. *Social Epistemology*, 14, 3–19.

Michael, M. and Brown, N. (2003) Dys-topias and dys-tropias: futures and performances in xenotransplantation. Paper presented at Annual British Sociology Association Conference, University of York, 11–13 April.

Michael, M. and Brown, N. (2004) The meat of the matter: grasping and judging xenotransplantation. *Public Understanding of Science*, 13, 379–397.

Michael, M. and Brown, N. (2005) Scientific citizenships: self-representations of xenotransplantation's publics. *Science as Culture*, 14, 38–57.

Michael, M. and Carter, S. (2001) The facts about fictions and vice versa: public understanding of human genetics. *Science as Culture*, 10, 5–32.

Millar, J. and Schwarz, M. (1998) Introduction: Speed is a vehicle. In J. Millar and M. Schwarz (eds) *Speed – Visions of an Accelerated Age.* London: The Photographers' Gallery and the Trustees of the Whitechapel Art Gallery.

Miller, D. (1998) Why some things matter. In D. Miller, (ed.) *Material Cultures*. London: UCL Press, 3–21.

Miller, D. (2001) *The Dialectics of Shopping*. Chicago, IL: Chicago University Press.

Mol, A. (2002) *The Body Multiple. Ontology in Medical Practice*. Durham, NC: Duke University Press.

Mol, A. and Law, J. (1994) Regions, networks and fluids: anaemia and social topology. *Social Studies of Science*, 24, 641–671.

Mol, A. and Law, J. (2004) Embodied action, enacted bodies: the example of hypoglycaemia. *Body and Society*, 10, 43–62.

Morley, D. (1992) *Television, Audiences and Cultural Studies*. London: Routledge.

Mulkay, M. (1997) *The Embryo Research Debate: The Science and Politics of Reproduction*. Cambridge: Cambridge University Press.

Muri, A. (2003) Of shit and the soul: tropes of cybernetic disembodiment in contemporary culture. *Body and Society*, 9, 73–92.

Myerson, G. (2000) *Donna Haraway and GM Foods*. Cambridge: Icon Books.

Nelkin, D. and Lindee, M.S. (1996) *The DNA Mystique. The Gene as a Cultural Icon*. New York, NY: W.H. Freeman & Company.

Nettleton, S., Pleace, N., Burrows, R., Muncer, S. and Loader, B. (2002) The reality of social support. In S. Woolgar (ed.) *Virtual Society? Technology, Cyberbole, Reality*. Oxford: Oxford University Press, 176–188.

Nettleton, S. and Watson, J. (1998) The body in everyday life: an introduction. In S. Nettleton and J. Watson (eds) *The Body in Everyday Life*. London: Routledge, 1–23.

Normark, D. (no date) Tending to mobility: intensities of staying at a petrol station, at http://www.tii.se/mobility/Files/EPNormarkfinal.pdf (accessed 14 August 2005).

Novas, C. and Rose, N. (2000) Genetic risk and the birth of the somatic individual. *Economy and Society*, 29, 485–513.

Nowotny, H. (1994) *Time: The Modern and Postmodern Experience*. Cambridge: Polity.

Nowotny, H., Scott, P. and Gibbons, M. (2001) *Re-thinking Science: Knowledge and the Public in an Age of Uncertainty*. Cambridge: Polity.

O'Connell, J. (1993) Metrology: the creation of universality by the circulation of particulars. *Social Studies of Science*, 23, 129–173.

O'Connell, S. (1998) *The Car in British Society: Class, Gender and Motoring 1896-1939*. Manchester: Manchester University Press.

Osborne, T. and Rose, N. (1999) Do the social sciences create phenomena: the case of public opinion research. *British Journal of Sociology* 50, 367–396.

Oteri, J.S., Weinberg, M. and Pinales, M.S. (1982) Cross-examinination of chemists in drugs cases. In B. Barnes and D. Edge (eds) *Science in Context*. Buckingham: Open University Press, 250–259.

Penley, C. and Ross, A. (eds) (1991) *Technoculture*. Minneapolis, MN: University of Minnesota Press.

Philo, C. and Wilbert, C. (2000) Animal spaces, beastly places: an introduction. In C. Philo and C. Wilbert (eds) *Animal Spaces, Beastly Places: New Geographies of Human – Animal Relations*. London: Routledge, 1–34.

Pickering, A. (1995) *The Mangle of Practice: Time, Agency and Science*. Chicago and London: University of Chicago Press.

Pickstone, J.V. (2000) *Ways of Knowing: A New Science, Technology and Medicine*. Manchester: Manchester University Press.

Poster, M. (2002) Everyday (virtual) life. *New Literary History*, 33, 743–760.

Power, M. (1999) *The Audit Society*. Oxford: Oxford University Press.

Prigogine, I. and Stengers, I. (1984) *Order out of Chaos*. London: Fontana.

Prins, B. (1995) The ethics of hybrid subjects: feminist constructivism according to Donna Haraway. *Science, Technology and Human Values*, 20, 352–367.

Rabinow, P. (1996) *Making PCR*. Chicago, IL: Chicago University Press.

Rappert, B. (1999) Rationalising the future? Foresight in science and technology policy co-ordination. *Futures*, 31, 527–545.

Rheinberger, H.-J. (1997) *Toward a History of Epistemic Things: Synthesizing Proteins in the Test Tube*. Palo Alto, CA: Stanford University Press.

Rose, N. (1996) *Inventing Our Selves: Psychology, Power and Personhood*. Cambridge: Cambridge University Press.

Rose, N. (1999) *Powers of Freedom*. Cambridge: Cambridge University Press.

Rose, N. (2005) Will Biomedicine Transform Society? The Political, Economic, Social and Personal Impact of Medical Advances in the Twenty-first Century. Clifford Barclay Lecture, London School of Economics and Political Science.

Rosen, P. (1993) The social construction of mountain bikes: technology and postmodernity in the cycle industry. *Social Studies of Science*, 23, 479–513.

Rosen, P. (2002) *Framing Production: Technology, Culture and Change in the British Bicycle Industry*. Cambridge, MA: MIT Press.

Rosengren, A. (1994) Some notes on the male motoring world in a Swedish community. In K.H. Sorensen (ed.) *The Car and its Environments*. Brussels: ECSC-EEC-EAEC.

Royal Society of London (1985) *The Public Understanding of Science*. London: The Royal Society.

Saunders, P. (1993) Citizenship in a liberal society. In B.S. Turner (ed.) *Citizenship and Social Theory*. London: Sage, 57–90.

Scheper-Hughes, N. (2004) Parts unknown: undercover ethnography of the organs-trafficking underworld. *Ethnography*, 5, 29–73.

Schlegoff, E. and Sacks, H. (1973) Opening up closings. *Semiotica*, 7, 289–327.

Schutz, A. (1967) *Phenomenology of the Social World*. London: Heinemann.

Seigworth, G.J. (2000) Banality for cultural studies. *Cultural Studies*, 14, 227–268.

Sellen, A.J. and Harper, R.H.R. (2003) *The Myth of the Paperless Office*. Cambridge, MA: MIT Press.

Serres, M. (1982a) *Hermes: Literature, Science, Philosophy*. Baltimore, MD: Johns Hopkins University Press.

Serres, M. (1982b) *The Parasite*. Baltimore, MD: Johns Hopkins University Press.

Serres, M. (1991) *Rome: The Book of Foundations*. Stanford, CA: Stanford University Press.

Serres, M. (1995a) *The Natural Contract*. Ann Arbor, MI: Michigan University Press.

Serres, M. (1995b) *Genesis*. Ann Arbor, MI: Michigan University Press.

Serres, M. (2003) The science of relations: an interview. *Angelaki: Journal of Theoretical Humanities*, 8, 227–238.

Serres, M. and Latour, B. (1995) *Conversations on Science, Culture and Time*. Ann Arbor, MI: Michigan University Press.

Shapin, S. (1991) Science and the public. In R.C. Olby, G.N. Cantor, J.R.R. Christie and M.J.S. Hodge (eds) *Companion to the History of Modern Science*. London: Routledge & Kegan Paul, 990–1007.

Sheller, M. and Urry, J. (2000) The city and the car. *International Journal of Urban and Regional Research*, 24, 737–757.

Shields, R. (1999) *Lefebvre, Love and Struggle: Spatial Dialectics*. London: Routledge.

Shilling, C. (1993) *The Body and Social Theory*. London: Sage.

Shove, E. (2003) *Comfort, Cleanliness and Convenience*. Oxford: Berg.

Shove, E. and Southerton. D. (2000) Defrosting the narrative: from novelty to convenience. *Journal of Material Culture*, 5, 301–319.

Silva, E.B. (2000) The cook, the cooker and the gendering of the kitchen. *Sociological Review*, 48, 612–628.

Silva, E.B. (2004) Materials and morals: families and technologies in everyday life. In E.B. Silva and T. Bennett (eds) *Contemporary Culture and Everyday Life*. Durham: sociologypress, 52–70.

Silva, E.B. and Bennett, T. (eds) (2004) *Contemporary Culture and Everyday Life*. Durham: sociologypress.

Silverstone, R. (1994) *Television and Everyday Life*. London: Routledge.

Silverstone, R. and Hirsch, E. (eds) (1992) *Consuming Technologies*. London: Routledge.

Silverstone, R., Hirsch, E. and Morley, D. (1992) Information and communication technologies and the moral economy of the household. In R. Silverstone and E. Hirsch (eds) *Consuming Technologies*. London: Routledge, 15–31.

Simmel, G. (1979) The metropolis and mental life. In D. Frisby and M. Featherstone (eds) *Simmel on Culture*. London: Sage, 174–185.

Singleton, V. (1993) *Science, Women and Ambivalence: An Actor Network Analysis of the Cervical Screening Programme*. Unpublished PhD thesis, Lancaster University.

Singleton, V. and Michael, M. (1993) Actor-networks and ambivalence: general practitioners in the cervical screening programme. *Social Studies of Science*, 23, 227–264.

Slater, D. (1997) *Consumer Culture and Modernity*. Cambridge: Polity.

Smith, D.E. (1988) *The Everyday World as Problematic: a Feminist Sociology*. Boston, MA: Northeastern University Press.

Sorensen, K. (2004) Cultural politics of technology: combining critical and constructionist interventions. *Science Technology and Human Values*, 29, 184–190.

Stenoien, J.M. (1994) Controlling the car: a regime change in the political understanding of traffic risk in Norway., In K.H. Sorensen (ed.) *The Car and its Environments*. Brussels: ECSC-EEC-EAEC.

Stzompka, P. (1999) *Trust: A Sociological Theory*. Cambridge: Cambridge University Press.

Synnott, A. (1993) *The Body Social: Symbolism, Self and Society*. London: Routledge.

Thomas, K. (1984) *Man and the Natural World*. Harmondsworth: Penguin.

Thrift, N. (2004) Driving in the city. *Theory, Culture and Society*, 21, 41–59.

Traweek, S. (1988) *Life Times and Beamtimes. The World of High Energy Physicists*. Cambridge, MA: Harvard University Press.

Tulloch, J. and Lupton, D. (2003) *Risk and Everyday Life*. London: Sage.

Turner, B. (1996) *The Body and Society* (2nd edn). London: Sage.

Turner, J. (1996) *Understanding the Public Understanding of Biotechnology in Late Modernity*. Unpublished PhD thesis, Lancaster University.

Turney, J. (1998) *Frankenstein's Footsteps: Science, Genetics and Popular Culture*. New Haven, CT: Yale University Press.

Turney, J. (2000) *Frankenstein's Footsteps: Science, Genetics and Popular Culture*. New Haven, CT: Yale University Press.

Urry, J. (1990) *The Tourist Gaze*. London: Sage.

Urry, J. (2000) *Sociology Beyond Societies*. London: Routledge.

Urry, J. (2003) *Global Complexity*. Cambridge: Polity.

Urry, J. (2004) The 'system' of automobility. *Theory, Culture & Society*, 21, 25–39.

Vaneigem, R. (1983) *The Revolution of Everyday Life*. London: Aldgate Press.

Waldby, C. (2000) *The Visible Human Project: Informatic Bodies and Posthuman Medicine*. London: Routledge.

Wardell, C. (2001) *Facts, Fictions and Futures: Towards a Cultural Understanding of the Public Understanding of Science*. Unpublished PhD thesis, Goldsmiths College, University of London.

Warner, M. (2002) *Publics and Counterpublics*. New York, NY: Zone Books.

Watkins, H. (2003) Fridge stories: three geographies of the domestic refrigerator, at http://www.ifs.tu-darmstadt.de/fileadmin/gradkoll//Publikationen/transformingspaces.html.

Weldon, S. (1998) *Runway Rhetorics and Networking with Nature: A Study of Scientific Expertise in Environmental Impact Assessment*. Unpublished PhD thesis, School of Independent Studies, Lancaster University.

Whatmore, S. (2002) *Hybrid Geographies*. London: Sage.

Whitehead, A.N. (1925) *Science and the Modern World*. New York, NY: Free Press.

Whitehead, A.N. (1929) *Process and Reality*. Cambridge: Cambridge University Press.

Whitely, R. (1984) *The Intellectual and Social Organization of the Sciences*. Oxford: Clarendon.

Williams, R. (1973) *The Country and the City*. London: Chatto & Windus.

Williams, R. (1988) *Keywords*. London: Fontana.

Williams, S.J. (2002) Sleep and health: sociological reflections on the dormant society. *Health*, 6, 173–200.

Williams, S.J. and Boden, S. (2004) Consumed with sleep? Dormant bodies in consumer culture. *Sociological Research Online*, 9, at http://www.socresonline.org.uk/9/2/williams.html.

Winner, L. (1988) *The Whale and the Reactor: A Search for Limits in an Age of High Technology*. Chicago, IL: University of Chicago Press.

Wolch, J. and Emel, J. (eds) (1998) *Animal Geographies: Place, Politics and Identity in the Nature–Culture Borderlands*. London: Verso.

Woolgar, S. (2002) Five rules of virtuality. In S. Woolgar (ed.) *Virtual Society? Technology, Cyberbole, Reality*. Oxford: Oxford University Press, 1–22.

Wyatt, S., Thomas, G. and Terranova, T. (2002) They came, they surfed, they went back to the beach: conceptualizing use and non-use of the internet. In S. Woolgar (ed.) *Virtual Society? Technology, Cyberbole, Reality*. Oxford: Oxford University Press, 23–40.

Wynne, B.E. (1991) Knowledges in context. *Science, Technology and Human Values*, 16, 111–121.

Wynne, B.E. (1992) Misunderstood misunderstanding: social identities and public uptake of science. *Public Understanding of Science*, 1, 281–304.

Wynne, B.E. (1995) The public understanding of science. In S. Jasanoff, G.E. Markle, J.C. Peterson and T. Pinch (eds) *Handbook of Science and Technology Studies*. Thousand Oaks, CA: Sage, 361–388.

Wynne, B.E. (1996) May the sheep safely graze? A reflexive view of the expert–lay divide. In S. Lash, B. Szerszynski and B. Wynne (eds) *Risk, Environment and Modernity*. London: Sage, 44–83.

Wynne, B.E. (2001) Creating public alienation: expert cultures of risk and ethics on GMOs. *Science as Culture*, 10, 445–482.

Wynne, B.E. (2003) Seasick on the third wave? Subverting the hegemony of pro-positionalism: response to Collins and Evans (2002). *Social Studies of Science*, 33, 401–417.

Index